Rinaldo B. Schinazi

Probability
with
Statistical Applications

Birkhäuser
Boston • Basel • Berlin

Rinaldo B. Schinazi
Department of Mathematics
University of Colorado
Colorado Springs, CO 80933-7150
U.S.A.

Library of Congress Cataloging-in-Publication Data

A CIP catalogue record for this book is available from the Library of Congress,
Washington D.C., USA.

AMS Subject Classifications: 60Axx, 62Exx, 60Fxx, 62Cxx, 62Jxx

Printed on acid-free paper
© 2001 Birkhäuser Boston

Birkhäuser

ISBN 0-8176-4247-1   SPIN 10838837
ISBN 3-7643-4247-1

Reformatted from author's files in LaTeX2ε by TeXniques, Inc., Cambridge, MA
Printed and bound by Hamilton Printing Company, Rensselaer, NY
Printed in the United States of America

9 8 7 6 5 4 3 2 1

To Ted Cox, teacher and friend

Preface

This book is intended as a one-semester first course in probability and statistics, requiring only a knowledge of calculus. It will be useful for students majoring in a number of disciplines: for example, biology, computer science, electrical engineering, mathematics, and physics.

Many good texts in probability and statistics are intended for a one-year course and consist of a large number of topics. In this book, the number of topics is drastically reduced. We concentrate instead on several important concepts that every student should understand and be able to apply in an interesting and useful way. Thus statistics is introduced at an early stage.

The presentation focuses on topics in probability and statistics and tries to minimize the difficulties students often have with calculus. Theory therefore is kept to a minimum and interesting examples are provided throughout. Chapter 1 contains the basic rules of probability and conditional probability with some interesting applications such as Bayes' rule and the birthday problem. In Chapter 2 discrete and continuous random variables, expectation and variance are introduced. This chapter is mostly computational with a few probability concepts and many applications of calculus. In Chapters 3 and 4 we get to the heart of the subject: binomial distribution, normal approximation of the binomial, Poisson distribution, the Law of Large Numbers and the Central Limit Theorem. We also cover the Poisson approximation of the binomial (in a nonstandard way) and the Poisson scattering theorem.

In Chapter 5 we apply some of the concepts of the preceding chapters to introduce statistics. We cover confidence intervals and hypothesis testing for large samples, and we also introduce student tests to deal with small samples and a nonparametric test. Finally, we test independence and goodness-of-fit by using Chi-square tests. Chapter 6 is a short introduction to linear regression. Chapters 7 and 8 rely heavily on the calculus of one and several variables to study sums of random variables (via moment generating functions), transformations of random variables (using distribution functions) and transformations of random vectors. In Chapter 8 we prove a number of facts regarding expectation, variance and covariance that are used throughout the book. We also prove facts about normal samples that are useful in statistics.

There are at least two ways to use this book for a one-semester course. Both should cover the first four chapters. Then one may choose to do some statistical applications as in Chapters 5 and 6, or one may concentrate on probability and cover Chapters 7 and 8. The graph below shows how the different chapters are related to one another.

I would like to thank my colleague Sarbarish Chakravarty for using my lecture notes in his class and for his constructive feedback. I would also like to thank an anonymous reviewer for his (or her) encouraging words and for suggestions that improved the presentation of the book.

Contents

Probability
with
Statistical Applications

1

Probability Space

1.1 The Axioms of Probability

The study of probability is concerned with the mathematical analysis of random experiments such as tossing a coin, rolling a die or playing at the lottery. Each time we perform a random experiment there are a number of possible outcomes. We first define the notions of sample space and event.

> **Sample Space and Events**
>
> The sample space Ω of a random experiment is the collection of all possible outcomes of the random experiment.
>
> An event is any subset of Ω.

Example 1. Toss a coin. There are only two possible outcomes and the sample space is $\Omega = \{H, T\}$. The event $A = \{H\}$ is equivalent to the event 'the outcome was heads'.

Example 2. Roll a die. This time the sample space is $\Omega = \{1, 2, 3, 4, 5, 6\}$. The event $B = \{1, 3, 5\}$ is equivalent to the event 'the die showed an odd face'.

Example 3. The birthday of someone. The sample space has 365 points (ignoring leap years).

Example 4. We count the number of rolls until we get a 6. Here $\Omega = \{1, 2, \ldots\}$. That is, the sample space consists of all strictly positive integers. Note that this sample space has infinitely many points.

We define next some useful relations among events.

If A is an event included in the sample space Ω, then the event consisting of all the points of Ω not included in A is called the complement of A and is denoted by A^c.

Assume that A and B are two events; then the intersection of A and B is the set of points that are both in A and B. The intersection of A and B is denoted by AB.

If A and B are two events, then the union of A and B is the set of points that are in A or in B (they may be in both). The union of A and B is denoted by $A \cup B$.

The empty set is denoted by $\emptyset$. Two events are said to be disjoint or mutually exclusive if

$$AB = \emptyset.$$

Example 5. Let A be the event that a student is female, B the event that a student takes French and C the event that a student takes calculus.

What is the event 'a student is female and takes calculus'? We want both A and C so the event is AC.

What is the event 'a student does not attend calculus'? We want everything not in C so the event is C^c.

What is the event 'a student takes French or calculus'? We want everything in A and everything in B so the event is $A \cup B$.

We now state several important set theory identities.

Set Theory Identities

Let A and B be two events. Then:

$$(A \cup B)^c = A^c B^c.$$

$$(A \cap B)^c = A^c \cup B^c.$$

$$A = AB \cup AB^c.$$

The identities above are not difficult to establish. For instance, for the first one x belongs to $(A \cup B)^c$ if and only if x does not belong to $A \cup B$; this in turn is equivalent to x not belonging to A *and* not belonging to B, which is equivalent to x belonging to A^c and to B^c and thus to $A^c B^c$.

We now give the Axioms (rules) of probability.

Axioms of Probability

(i) A probability P is a function defined on the set of events of a sample space Ω with values in $[0,1]$.

(ii) $P(\Omega) = 1$.

(iii) For a finite or infinite sequence of disjoint events A_i,
$$P(\cup A_i) = \sum P(A_i).$$

Consequences

C1. If $AB = \emptyset$, then by (iii)

$$P(A \cup B) = P(A) + P(B).$$

C2. Using C1 with $B = A^c$ we get

$$P(A^c) = 1 - P(A).$$

C3. Using $A = \Omega$ in C2 we get

$$P(\emptyset) = 0.$$

C4. Using that $AB \cap AB^c = \emptyset$ and that $A = AB \cup AB^c$ we get by (iii) that

$$P(A) = P(AB) + P(AB^c).$$

C5. Using that $A \cup B = AB^c \cup B$ and that the last two events are disjoint we get by C1 that $P(A \cup B) = P(AB^c) + P(B)$. Now using C4 we know that $P(AB^c) = P(A) - P(AB)$. Thus, for any two events A and B (in particular they do not need to be disjoint) we have

Union of Two Events

$$P(A \cup B) = P(A) + P(B) - P(AB).$$

Example 6. We pick at random a person in a certain population. Let A be the event that the person selected attends college. Let B be the event that the person selected speaks French. Assume that the proportion of persons attending college and speaking French in the population are 0.1 and 0.02, respectively. Then it makes sense to define $P(A) = 0.1$ and $P(B) = 0.02$. Assume also that the proportion of people attending college and speaking French is 0.01. That is, P(AB)=0.01.

What is the probability that a person picked at random does not attend college? This is the event A^c. By C2 we have

$$P(A^c) = 1 - P(A) = 0.9.$$

What is the probability that a person picked at random speaks French or attends college?

This is the event $A \cup B$. By C5 we have

$$P(A \cup B) = P(A) + P(B) - P(AB) = 0.1 + 0.02 - 0.01 = 0.11.$$

What is the probability that a person speaks French and does not attend college? This is the event $A^c B$. According to C4 we have

$$P(A^c B) = P(B) - P(AB) = 0.02 - 0.01 = 0.01.$$

Equally likely outcomes

We start by considering an example.

Example 7. Roll a fair die. Then $\Omega = \{1, 2, 3, 4, 5, 6\}$. Since all outcomes are equally likely it makes sense to define

$$P(i) = 1/6 \text{ for } i = 1, \ldots, 6.$$

What is the interpretation of the statement $P(1) = 1/6$? If we roll the die many times the frequency of observed 1s (that is, the observed number of 1s divided by the total number of rolls) should be close to 1/6.

What is the probability of the die showing an odd face? By axiom of probability (iii),

$$P(odd) = P(\{1, 3, 5\}) = P(\{1\}) + P(\{3\}) + P(\{5\}) = 3/6.$$

More generally, we have the following.

Equally Likely Outcomes

Consider a finite sample space Ω with finitely many outcomes assumed to be equally likely. Let $|A|$ be the number of elements in A. Then we define

$$P(A) = \frac{|A|}{|\Omega|}$$

for every event A.

It is easy to check that P defined by the formula above satisfies the three axioms of probability and thus is a probability on Ω.

Example 8. Toss two fair coins. This time we have four equally likely outcomes; $\Omega = \{HH, HT, TH, TT\}$.

$$P(\text{ at least 1 head}) = \frac{|\{HH, HT, TH\}|}{4} = 3/4.$$

Example 9. Roll two dice. What is the probability that the sum is 11? The most natural sample space is all the possible sums: so all integers from 2 to 12. But these outcomes are not equally likely so it is not a good choice. Instead we pick for Ω the collection of all ordered pairs: $\{(1, 1), (1, 2), \ldots, (2, 1), (2, 2), \ldots, (6, 5), (6, 6)\}$. There are 36 equally likely outcomes in Ω.

$$P(\text{ sum is 11}) = \frac{|\{(5, 6), (6, 5)\}|}{36} = 2/36.$$

Example 10. Roll two dice. What is the probability that the sum of the two dice is 4 or more? It is quicker to compute the probability that the sum is 3 or less, which is the complement of the event we want.

$$P(\text{ sum is 3 or less}) = \frac{|\{(1, 1), (1, 2), (2, 1)\}|}{36} = 3/36.$$

Therefore,

$$P(\text{ sum is 4 or more}) = 1 - 3/36 = 33/36.$$

Exercises

1. Let A be the event that a person attends college and B be the event that a person speaks French. Using intersections, unions or complements describe the following events.
   (a) A person does not speak French.
   (b) A person speaks French and does not attend college.
   (c) A person is either in college or speaks French.
   (d) A person is either in college or speaks French but not both.

2. Let A and B be events such that $P(A) = 0.6$, $P(B) = 0.3$ and $P(AB) = 0.1$.
   (a) Find the probability that A or B occurs.
   (b) Find the probability that exactly one of A or B occurs.
   (c) Find the probability that at most one of the two events A and B occurs.
   (d) Find the probability that neither A nor B occurs.

3. Toss three fair coins.

(a) What is the probability of having at least one head?

(b) What is the probability of having exactly one head?

4. Roll two fair dice.

(a) What is the probability that they do not show the same face?

(b) What is the probability that the sum is 7?

(c) What is the probability that the maximum of the two faces is at least 3?

5. In a college it is estimated that 1/4 of the students drink, 1/8 of the students smoke and 1/10 smoke and drink. Picking a student at random,

(a) what is the probability that the student does not drink nor smoke?

(b) what is the probability that a student smokes or drinks?

6. A roulette has 38 pockets, 18 are red, 18 are black and 2 are green. I bet on red, you bet on black.

(a) What is the probability that I win?

(b) What is the probability that at least one of us wins?

(c) What is the probability that at least one of us loses?

7. Roll 3 dice.

(a) What is the probability that you get three 7s?

(b) What is the probability that you get a triplet?

(c) What is the probability that you get a pair?

8. I buy many items at a grocery store. What is the probability that the bill will be a whole number?

9. If $A \subset B$, show that $P(A^c B) = P(B) - P(A)$.

10. Show that for any three events A,B,C we have

$$P(A \cup B \cup C) = P(A) + P(B) + P(C) - P(AB) - P(AC) - P(BC) + P(ABC).$$

Can you guess what the formula is for the union of four events?

1.2   Conditional Probabilities and Bayes' Formula

Example 1. Roll two dice successively and observe the sum. As we observed before we should take for our sample space the 36 ordered pairs. Let A be the event 'the sum is 11'. Since all the outcomes are equally likely we have that

$$P(A) = \frac{|\{(5, 6), (6, 5)\}|}{36} = 1/18.$$

Let B be the event 'the first die shows a 6'. We are now interested in the following question: if we observe the first die and it shows a 6, how does this affect the probability of observing a sum of 11? In other words, given B, what is the probability of A? Observe that for this question our sample space is B. The notation for the preceding probability is

$$P(A|B)$$

and is read 'probability of A given B'. Given that the first die shows a 6 there is only one possibility for the sum to be 11. The second die needs to show 5. The probability of this event is 1/6. Thus,

$$P(A|B) = 1/6.$$

More generally, we have the following definition.

Conditional Probability

The probability of A given B is defined by

$$P(A|B) = \frac{P(AB)}{P(B)}.$$

In the case of equally likely outcomes the formula becomes

$$P(A|B) = \frac{|AB|}{|B|}.$$

By using the definition above it is easy to see that the rules of probability apply to conditional probabilities. In particular,

$$P(A \cup B|C) = P(A|C) + P(B|C) \text{ if } A \text{ and } B \text{ are disjoint}$$

and

$$P(A^c|B) = 1 - P(A|B).$$

Example 2. We pick at random a person in a certain population. Let A be the event that the person selected attends college. Let B be the event that the person selected speaks French. Assume that the proportion of persons attending college and speaking French in the population are 0.1 and 0.02, respectively. Assume also that the proportion of people attending college and speaking French is 0.01. Given that the person we picked speaks French, what is the probability that this person attends college? We want

$$P(A|B) = \frac{P(AB)}{P(B)} = 0.01/0.02 = 1/2.$$

Given that the selected person attends college, what is the probability that this person speaks French? This time we want

$$P(B|A) = \frac{P(AB)}{P(A)} = 0.01/0.1 = 0.1.$$

Given that the selected person attends college, what is the probability that this person does not speak French?

$$P(B^c|A) = 1 - P(B|A) = 1 - 0.1 = 0.9.$$

The previous two examples show how to compute conditional probabilities by using unconditional probabilities. In many situations, as we are going to see next, it is the reverse that is useful: the conditional probabilities are easy to compute and we use them to compute unconditional probabilities. Note first that the definition of conditional probability is equivalent to the following rule.

Multiplication Rule

$$P(AB) = P(A|B)P(B).$$

Example 3. A factory has an old (O) and a new (N) machine. The new machine produces 70% of the products and 1% of these products are defective. The old machine produces the remaining 30% of the products and of those 5% are defective. All products are randomly mixed. What is the probability that a product picked at random is defective and produced by the new machine?

Let D be the event that the product picked at random is defective. Note that the following probabilities are given.

$$P(N) = 0.7, \quad P(O) = 0.3, \quad P(D|N) = 0.01 \text{ and } P(D|O) = 0.05.$$

We want the probability of DN. By the multiplication rule we have

$$P(DN) = P(D|N)P(N) = 0.01(0.7) = 0.007.$$

Assume now that we are interested in the probability that a product picked at random is defective. We can write

$$P(D) = P(DN) + P(DO).$$

That is, a defective product may come from the new or the old machine. Now we use the multiplication rule twice to get

$$P(D) = P(D|N)P(N) + P(D|O)P(O) = 0.01(0.7) + 0.05(0.3) = 0.022.$$

That is, we get the overall defective proportion by averaging the defective proportion for each machine. This is a very useful way of proceeding and we now state the rule in its general form.

Rule of Average

For any events A and B we have

$$P(A) = P(A|B)P(B) + P(A|B^c)P(B^c).$$

More generally, if the events $B_1, B_2, \ldots, B_n$ are disjoint and if their union is the whole sample space Ω, then

$$P(A) = P(A|B_1)P(B_1) + P(A|B_2)P(B_2) + \cdots + P(A|B_n)P(B_n).$$

We now apply the rule of average to another example.

Example 4. We have three boxes labeled 1, 2 and 3. Box 1 has one white ball and two black balls, Box 2 has two white balls and one black ball and Box 3 has three white balls. One of the three boxes is picked at random and then a ball is picked from this box. What is the probability that the ball picked is white?

Let W be the event 'the ball picked is white'. We use the rule of average and get

$$P(W) = P(W|1)P(1) + P(W|2)P(2) + P(W|3)P(3).$$

The conditional probabilities above are easy to compute. We have

$$P(W|1) = 1/3, \quad P(W|2) = 2/3, \quad P(W|3) = 1.$$

Thus,

$$P(W) = 1/3 \times 1/3 + 2/3 \times 1/3 + 1 \times 1/3 = 2/3.$$

As we have just seen, the conditional probability $P(W|1)$ is easy to compute. What about $P(1|W)$? That is, given that we picked a white ball what is the probability that it came from box 1?

In order to answer this question we start by using the definition of conditional probability.

$$P(1|W) = \frac{P(1W)}{P(W)}.$$

Now we use the multiplication rule for the numerator and the average rule for the denominator. We get

$$P(1|W) = \frac{P(W|1)P(1)}{P(W|1)P(1) + P(W|2)P(2) + P(W|3)P(3)}.$$

Numerically, we have

$$P(1|W) = \frac{1/3 \times 1/3}{2/3} = 1/6.$$

Note that $P(1|W)$ is twice less likely than $P(1)$. That is, given that the ball drawn is white, box 1 is less likely to have been picked than boxes 2 and 3. Since box 1 has fewer white balls than the other boxes this is not surprising. The preceding method applies each time we want the conditional probability $P(A|B)$, but what is readily available is the conditional probability $P(B|A)$. We now state the general form of this useful formula.

Bayes' Formula

For any events A and B we have

$$P(B|A) = \frac{P(A|B)P(B)}{P(A|B)P(B) + P(A|B^c)P(B^c)}.$$

More generally, if the events $B_1, B_2, \ldots, B_n$ are disjoint and if their union is the whole sample space Ω, then for every $i = 1, \ldots, n$

$$P(B_i|A) = \frac{P(A|B_i)P(B_i)}{P(A|B_1)P(B_1) + P(A|B_2)P(B_2) + \cdots + P(A|B_n)P(B_n)}.$$

As observed in Example 4, Bayes' formula is an easy consequence of the definition of conditional probabilities and the rule of average. Rather than memorizing it the reader should become familiar with the way to derive it. Next we give another example of the use of Bayes' rule.

Example 5. It is estimated that 10% of the population has a certain disease. A diagnostic test is available but is not perfect. The probability that a healthy person will be diagnosed with the disease is 5%. The probability that someone with the disease will be diagnosed as healthy is 1%. Given that a person picked at random is diagnosed with the disease, what is the probability that this person actually has the disease?

Let D be the event that the person has the disease, $+$ be the event that the person is diagnosed as having the disease, $-$ the event that the person is diagnosed as not having the disease. We are asked to compute the conditional probability $P(D|+)$. Note that $P(+|D) = 1 - 0.01 = .99$; but $P(D|+)$ is not as readily available so we use Bayes' formula.

$$P(D|+) = \frac{P(D+)}{P(+)} = \frac{P(+|D)P(D)}{P(+|D)P(D) + P(+|D^c)P(D^c)}.$$

We know that $P(D) = 0.1$ so $P(D^c) = 0.9$. As observed before, $P(+|D) = .99$ and $P(+|D^c) = 0.05$. Thus,

$$P(D|+) = \frac{.99 \times .1}{.99 \times .1 + .05 \times .9} = .69.$$

So the probability that the person has been misdiagnosed positive is .31 which is quite high! This is so because the proportion of people with the disease is comparable to the proportion of misdiagnoses.

Symmetry

It is sometimes possible to avoid lengthy computations by invoking symmetry in a problem. We give next such an example.

Example 6. You are dealt two cards from a deck of 52 cards. What is the probability that the second card is black?

One way to answer the preceding question is to condition on whether the first card is black. Let B and R be the events 'the first card is black' and the 'first card is red', respectively. Let A be the event 'the second card is black'. We have

$$P(A) = P(AR) + P(AB) = P(A|R)P(R) + P(A|B)P(B)$$
$$= (26/51)(1/2) + (25/51)(1/2) = 1/2.$$

Now we show how a symmetry argument yields this result. By symmetry we have

$$P(\text{ the second card is red}) = P(\text{ the second card is black}).$$

Since

$$P(\text{ the second card is red}) + P(\text{ the second card is black}) = 1,$$

we get that
$$P(\text{ the second card is black}) = 1/2.$$

Exercises

1. Consider the student population in a college campus. Assume that 55% of the students are female. Assume that 20% of the males drink and 10% of the females drink.

(a) Pick a female student at random; what is the probability that she does not drink?

(b) Pick a student at random; what is the probability that the student does not drink?

(c) Pick a student at random; what is the probability that this student is male and drinks?

2. A company has two factories A and B. Assume that factory A produces 80% of the products and B the remaining 20%. The proportion of defectives are 0.05 for A and 0.01 for B.

(a) What is the probability that a product picked at random comes from A and is not defective?

(b) What is the probability that a product picked at random is defective?

3. Consider two boxes labeled 1 and 2. In box 1 there are two black balls and three white balls. In box 2 there are three black balls and two white balls. We pick box 1 with probability 1/3 and box 2 with probability 2/3. Then we draw a ball in the box we picked.

(a) Given that we pick box 2, what is the probability of drawing a white ball?

(b) Given that we draw a white ball, what is the probability that we picked box 1?

(c) What is the probability of picking a black ball?

4. Consider an electronic circuit with components $C1$ and $C2$. The probability that $C1$ fails is 0.1. If $C1$ fails, the probability that $C2$ fails is 0.15. If $C1$ works, the probability that $C2$ fails is 0.05.

(a) What is the probability that both components fail?

(b) What is the probability that at least one component works?

(c) What is the probability that $C2$ works?

5. Suppose five cards are dealt from a deck of 52 cards.

(a) What is the probability that the second card is a queen?

(b) What is the probability that the fifth card is a heart?

6. Two cards are dealt from a deck of 52 cards. Given that the first card is red, what is the probability that the second card is a heart?

7. A factory tests all its products. The proportion of defective items is 0.01. The probability that the test will catch a defective product is 0.95. The test will also reject non-defective products with probability 0.01.

(a) Given that a product passes the test, what is the probability that it is defective?

(b) Given that the product does not pass the test, what is the probability that the product is defective?

8. Consider the following game. There are three balls in a box, two are white and one is black. You win the game if you pick a white ball. You draw a ball but you do not see the color of the ball. Then someone takes out of the box a white ball. So at this point there is only one ball left in the box. At this point the rules of the game allow you to switch your ball with the one remaining in the box. What is the best strategy: to switch balls or not? In order to decide, compute the probability of winning for each strategy.

9. Two cards are selected from a 52 cards deck. The two cards are said to form a blackjack if one of the cards is an ace and the other is either a 10, a Jack, a Queen or a King. What is the probability that two cards form a blackjack?

10. Two dice are rolled. Given that the sum is 9, what is the probability that at least one die landed on 6?

11. Assume that 1% of men and 0.01% of women are colorblind. A colorblind person is chosen at random. What is the probability that this person is a man?

12. Hemophilia is a genetic disease that is caused by a recessive gene on the X chromosome. A woman is said to be a carrier of the disease if she has the hemophilia gene on one X chromosome and the healthy gene on the other X chromosome. A woman carrier has probability 1/2 of transmitting the disease to each son, since a son will get an X chromosome from the mother and a Y chromosome from the father. Because of her family history a woman is thought to have a 50% chance of being a carrier before having children. Given that this woman has three healthy sons, what is the probability that she is a carrier?

1.3   Independent Events

We start with Example 4 of the preceding section.

Example 1. We have three boxes labeled 1, 2 and 3. Box 1 has one white ball and two black balls, Box 2 has two white balls and one black ball and Box 3 has three white balls. One of the three boxes is picked at random and then a ball is picked from this box. Given that we draw a white ball, what is the probability that we have picked box 1?

We have already computed this conditional probability and found it to be 1/6. On the other hand the (unconditional) probability of picking box 1 is 1/3. So the information that the ball drawn is white changes the probability of picking box 1. In this sense we say that the events $A = \{$ box 1 is picked $\}$ and $B = \{$ a white ball is drawn $\}$ are not independent. This leads to the following definition.

Independent Events

Two events A and B are said to be independent if

$$P(A|B) = P(A).$$

We have the following consequences from this definition.

C1. By using the definition of conditional probability we see that A and B are independent if and only if

$$P(AB) = P(A)P(B).$$

C2. If A and B are independent so are A and B^c. In order to see this write that

$$P(A) = P(AB) + P(AB^c).$$

By using C1 we get

$$P(AB^c) = P(A) - P(A)P(B) = P(A)(1 - P(B)) = P(A)P(B^c)$$

and this shows that A and B^c are independent.

C3. If A and B are independent so are A^c and B^c.

Example 2. Consider again the three boxes of Example 1 but this time we put the same number of white balls in each box. For instance, assume that each box has two white balls and one black ball. Are the events $A = \{$ box 1 is picked $\}$ and $B = \{$ a white ball is drawn $\}$ independent?

By Bayes' formula we have

$$P(A|B) = \frac{1/3 \times 2/3}{1/3 \times 2/3 + 1/3 \times 2/3 + 1/3 \times 2/3} = 1/3 = P(A).$$

So this time A and B are independent. This should be intuitively clear: this time the fact that the ball drawn is white does not yield additional information about which box was picked since all boxes have the same proportion of white balls.

Example 3. Assume that A and B are independent events such that $P(A) = 0.1$ and $P(B) = 0.3$. What is the probability that A or B occurs?

We want $P(A \cup B)$. Recall that

$$P(A \cup B) = P(A) + P(B) - P(AB).$$

By C1 we have

$$P(A \cup B) = 0.1 + 0.3 - 0.1 \times 0.3 = 0.37.$$

Example 4. Assume that A and B are independent; can they also be disjoint?

If A and B are disjoint, then $AB = \emptyset$. Thus, $P(AB) = 0$. However, if A and B are also independent, then

$$P(AB) = P(A)P(B) = 0.$$

Thus, $P(A) = 0$ or $P(B) = 0$. So if A and B are independent they may be disjoint if and only if one of these events has probability zero.

Example 5. Assume two components are in series as below.

Assume that each component fails independently of the other with probability 0.01. What is the probability that the circuit fails?

In order for the circuit to fail we must have that one of the two components fails. Let A be the event that the left component fails and B be the event that the right component fails. We want

$$P(A \cup B) = P(A) + P(B) - P(AB) = 0.01 + 0.01 - 0.0001 = 0.0199.$$

Example 6. Assume two components are in parallel as below.

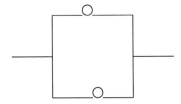

Assume they fail independently with probability 0.01. What is the probability that the circuit fails?

The circuit fails if both components fail.

$$P(AB) = P(A)P(B) = 0.0001.$$

As expected, the reliability of a parallel circuit is superior to the reliability of a series circuit. However, it is the independence assumption that greatly increases the reliability. The independence assumption may or may not be realistic.

Exercises

1. Assume that A and B are independent events with $P(A) = 0.2$ and $P(B) = 0.5$.
(a) What is the probability that exactly one of the events A and B occurs?
(b) What is the probability that neither A nor B occurs?
(c) What is the probability that at least one of the events A or B occurs?

2. Two cards are successively dealt from a deck of 52 cards. Let A be the event 'the first card is an ace' and B be event 'the second card is a spade'. Are these two events independent?

3. Two cards are successively dealt from a deck of 52 cards. Let A be the event 'the first card is an ace' and B be event 'the second card is an ace'. Are these two events independent?

4. Roll two dice. Let A be the event 'there is at least one 6' and B the event 'the sum is 7'. Are these two events independent?

5. Assume that the proportion of male students that drink is .2. Assume that there are 60% male students and 40% female students.
(a) Pick a student at random. What should the proportion of female drinkers be in order for the events 'the student is male' and 'the student drinks' be independent?
(b) Does your answer in (a) depend on the proportion of male students?

6. Show C3.

7. Assume that three components are as below.

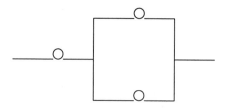

Assume that each component fails independently of the others with probability p_i, for $i = 1, 2, 3$. Find the probability that the circuit fails as a function of the p_i's.

8. Two components are in parallel. Assume that component 1 fails with probability p_1 given that component 2 is working. Assume that component 1 fails with probability sp_1 given that component 2 has failed. Similarly, component 2 fails with probability p_2 given that 1 is working and with probability sp_2 given that 1 has failed. Find the probability that the circuit fails.

9. Two cards are dealt from a 52 cards deck.
   (a) What is the probability of getting a pair?
   (b) What is the probability of getting two cards of the same suit?

1.4   Three or More Events

In this section we deal with probabilities involving several events. Our main tool is a generalization of the multiplication rule of Section 1.3. We now derive it for three events A, B and C. We start by using the multiplication rule for the two events AB and C.

$$P(ABC) = P(C|AB)P(AB).$$

We use the multiplication rule for A and B in the preceding equality. This yields

$$P(ABC) = P(C|AB)P(B|A)P(A) = P(A)P(B|A)P(C|AB).$$

The same computation can be done for an arbitrary number of events and yields the following.

Multiplication Rule for Three or More Events

Consider n events $A_1, A_2, \ldots, A_n$. The probability of the intersection of these n events can be written by using the conditional probabilities

$$P(A_1 A_2 \ldots A_n) = P(A_1)P(A_2|A_1)P(A_3|A_1 A_2) \ldots P(A_n|A_1 A_2 \ldots A_{n-1}).$$

We now apply this formula to several examples.

Example 1. Deal four cards from a deck of 52 cards. What is the probability of getting four aces?

Let A_1 be the event that the first card is an ace, let A_2 be the event that the second card is an ace and so on. We want to compute the probability of $A_1 A_2 A_3 A_4$. We use the multiplication rule above to get

$$P(A_1 A_2 A_3 A_4) = P(A_1) P(A_2|A_1) P(A_3|A_1 A_2) P(A_4|A_1 A_2 A_3).$$

The probability of A_1 is 4/52. Given that the first card is an ace the probability that the second card is an ace, is 3/51 and so on. Thus,

$$P(A_1 A_2 A_3 A_4) = 4/52 \times 3/51 \times 2/50 \times 1/49 = \frac{24}{6,497,400}.$$

A pretty slim chance to get four aces!

Example 2. We now deal with the famous birthday problem. Assume that there are 50 students in a class. What is the probability that at least two students have the same birthday?

It is easier to deal with the complement of this event. That is, we are going to compute the probability that all 50 students were born on different days. Assume that we are going through a list of the 50 birthdays in the class. Let B_2 be the event that the second birthday in the list is different from the first. Let B_3 be the event that the third birthday on the list is different from the first two. More generally, let B_i be the event that the ith birthday on the list is different from the first $i - 1$ birthdays for $i = 2, 3, \ldots, 50$. We want to compute the probability of $B_2 B_3 \ldots B_{50}$. By the multiplication rule we have

$$P(B_2 B_3 \ldots B_{50}) = P(B_2) P(B_3|B_2) P(B_4|B_2 B_3) \ldots P(B_{50}|B_2 B_3 \ldots B_{49}).$$

Ignoring the leap years, we assume that there are 365 days in a year. We also assume that all days are equally likely for birthdays. Note that $P(B_2) = 364/365$. Given that the first two birthdays are distinct, the third birthday has only 363 choices in order to be distinct from the first two. So $P(B_3|B_2) = 363/365$. The same reasoning shows that $P(B_4|B_3 B_2) = 362/365$. By doing the same type of computation for every term in the product above we get

$$P(B_2 B_3 \ldots B_{50}) = 364/365 \times 363/365 \times 362/365 \times \cdots \times 316/365.$$

The numerical computation gives a value of 0.96 for the probability of having at least two students having the same birthday in a class of 50! More generally, we have that

$$P(n \text{ people have } n \text{ distinct birthdays}) = \frac{364 \times 363 \times \cdots \times (365 - n + 1)}{365^{n-1}}.$$

The product above decreases rapidly to 0. If $n = 23$ we get that this product is about 0.50. For $n = 45$ it is about 0.05. Exercise 10 below will show how to approximate the product above by an exponential function. Note that if there are 365 people or more, then the probability of having 365 or more distinct birthdays is zero.

Independence

We now consider the independence property for several events. We have the following definition.

Independent Events

Three events A, B and C are said to be independent if the following conditions hold:
$$P(ABC) = P(A)P(B)P(C).$$

$$P(AB) = P(A)P(B), \quad P(AC) = P(A)P(C) \text{ and } P(BC) = P(B)P(C).$$

In general, n events are independent if for every integer k between 2 and n and any choice of k events (among the n we are considering) the probability of the intersection of these k events is the product of the probabilities of the k events.

The number of conditions to be checked grows rapidly with the number of events. It will be in general difficult to check that more than three events are independent. Typically, we will *assume* that events are independent (if that seems like a reasonable hypothesis) and then use the multiplication rule above to compute probabilities of interest. We illustrate this point next.

Example 3. Consider a class of 50 students. What is the probability that at least one of the students was born on December 25?

This is yet another case where it is easier to look at the complement of the event. We look at the list of birthdays in the class. Let A_i be the event that the ith student in the list was not born on December 25, for $1 \leq i \leq 50$. It is reasonable to assume that the A_i are independent: to know whether or not a certain student was born on December 25 does not give us additional information about the birthdays of other students (unless there are twins in the class and we assume that is not the case). By the independence assumption we have

$$P(A_1 A_2 \ldots A_{50}) = P(A_1)P(A_2) \ldots P(A_{50}).$$

Note that each A_i has probability 364/365. Thus,

$$P(A_1 A_2 \ldots A_{50}) = (364/365)^{50} = 0.87.$$

That is, the probability that at least one student in a class of 50 was born on a certain fixed day is about 0.13. The reader should compare this value with the value in Example 2.

Example 4. How many students should we have in a class in order to have at least one birthday on December 25 with probability at least 0.5?

Let n be the minimum number of students that satisfies the condition above. We use the events A_i, for $1 \leq i \leq n$, defined in Example 3. We want

$$P(A_1 A_2 \ldots A_n) \leq 0.5.$$

By independence we have that

$$(364/365)^n \leq 0.5.$$

We take logarithms on both sides of the inequality to get

$$n \ln(364/365) \leq \ln(0.5).$$

Recall that $\ln x < 0$ if $x < 1$. Thus,

$$n \geq \frac{\ln(0.5)}{\ln(364/365)}.$$

Numerically we get that n needs to be at least 253.

Exercises

1. Assume that three friends are randomly assigned to five classes. What is the probability that they are all in distinct classes?

2. Five cards are dealt from a 52 cards deck.
   (a) What is the probability that the five cards are all hearts?
   (b) What is the probability of a flush (all cards of the same suit)?

3. Roll five fair dice. What is the probability that at least two dice show the same face?

4. What is the probability of getting at least one 6 in ten rolls of a fair die?

5. Assume that the chance to win at the lottery with one ticket is 1/1,000,000. Assume that you buy one ticket per week. How many weeks should you play to have at least 0.5 probability of winning at least once?

6. Three electric components are in parallel. Each component fails independently of the others with probability p_i, $i = 1, 2, 3$. What is the probability that the circuit fails?

7. Three electric components are in series. Each component fails independently of the others with probability p_i, $i = 1, 2, 3$. What is the probability that the circuit fails?

8. Roll a die four times.
   (a) What is the probability of getting the same face four times?
   (b) What is the probability of getting the same face three times?

9. The probability of winning a certain game is $1/N$ for some fixed N. Show that you need to play the game approximately $\frac{2}{3}N$ times in order for the probability to win at least once be 0.5 or more. (Use that $\ln 2$ is approximately $2/3$ and that $\ln(1 - 1/N)$ is approximately $-1/N$ for N large).

10. In this exercise we are going to derive an approximate formula for the birthday problem (Example 2). Our starting point is that

$$p_n = P(n \text{ people have } n \text{ distinct birthdays}) = \frac{364 \times 363 \times \cdots \times (365 - n + 1)}{365^{n-1}}.$$

(a) Show that $\ln(p_n) = \ln(1-1/365)+\ln(1-2/365)+\cdots+\ln(1-(n-1)/365)$.

(b) Use that $\ln(1 - x)$ is approximately $-x$ for x near zero to show that $\ln(p_n)$ is approximately $-1/365 - 2/365 - \ldots - (n - 1)/365$.

(c) Show that $1 + 2 + 3 + \cdots + n = n(n + 1)/2$.

(d) Use (c) in (b) to show that $\ln(p_n)$ is approximately $\frac{-n(n-1)}{2\times365}$.

(e) Show that p_n is approximately

$$e^{\frac{-n(n-1)}{2\times365}}.$$

(f) Compute p_n for $n = 5, 10, 20, 30, 40, 50$ by using the exact formula and the approximation.

Review Exercises for Chapter 1

1. Assume that $P(A) = 0.1$ and $P(AB) = 0.05$.
(a) What is the probability that A occurs and B does not occur?
(b) What is the probability that A or B do not occur?

2. I draw one card from a deck of 52 cards. Let A be the event 'I draw a King' and let B be the event 'I draw a heart'. Are A and B independent?

3. Roll three dice. What is the probability of getting at least one 6?

4. I draw five cards from a deck of 52 cards.
(a) What is the probability that I get four Kings?
(b) What is the probability that I get four of a kind?

5. (a) I roll a die until the first 6 appears. What is the probability that I need six or more rolls?
(b) How many times should I roll the die so that I get at least one six with probability at least 0.9?

6. I draw five cards from a deck of 52 cards.
(a) What is the probability that I get no spade.
(b) What is the probability that I get no black cards?

7. I draw cards from a deck until I get a spade.

(a) What is the probability that I need exactly seven draws?

(b) Given that six or more draws are required, what is the probability that exactly seven draws are required?

8. Box 1 contains two red balls and three black balls. Box 2 contains six red balls and b black balls. We pick one of the two boxes at random and draw a ball from that box. Find b so that the color of the ball is independent of which box is picked.

9. 0s and 1s are sent down a communication channel. Assume that $P($ receive 0|transmit 0)=$P($ receive 1|transmit 1)=.99. Assuming that 0s and 1s are equally likely, what is the probability of a transmission error?

10. A student goes to class on a snowy day with probability 0.5 and on a non-snowy day with probability 0.8. Assume that 10% of the days in January are snowy. Given that the student was in class on January 28, what is the probability that it snowed that day?

11. One die is biased and the probability of a 6 is 1/2. The other die is fair. You pick one die at random and roll it. Given that you got a 6, what is the probability that you picked the biased die?

12. Consider a placement test for calculus. Assume that 80% of the students pass the placement test and that 70% of the students pass calculus. Experience has shown that 90% of the students who fail the placement test will fail calculus. What is the probability that a student who passes the placement test will pass calculus?

13. Consider a slot machine with three wheels, each marked with twenty symbols. On the central wheel, nine of 20 symbols are bells; on each of the left and right wheels there is one bell. In order to win the jackpot one has to get three bells. Assume that the three wheels spin independently and that every symbol is equally likely.

(a) What is the probability of hitting the jackpot?

(b) What is the probability of getting exactly two bells?

(c) Can you think of another distribution of bells that does not change the probability of hitting the jackpot but decreases the probability of getting exactly two bells?

14. Assume that A, B and C are independent events with probabilities 1/10, 1/5 and 1/2, respectively.

(a) Compute $P(ABC)$.

(b) Compute $P(A \cup B \cup C)$.

2
Random Variables

2.1 Discrete Random Variables

We start with an example.

Example 1. Toss two fair coins. Let X be the number of heads. X is a function from the sample space $\Omega = \{HH, HT, TH, TT\}$ into the set $\{0, 1, 2\}$. The *distribution* of X is given by the following table.

k	0	1	2
$P(X = k)$	1/4	1/2	1/4

More generally, we have the following definition.

Discrete Random Variables

A discrete random variable is a function from a sample space Ω into a countable set (usually the positive integers). The distribution of a random variable X is the sequence of probabilities $P(X = k)$ for all k in the range of X. We must have

$$P(X = k) \geq 0 \text{ for every } k \text{ and } \sum_k P(X = k) = 1.$$

The term *discrete* refers to the fact that the random variables, in this section, take values in countable sets. The next section deals with *continuous* random variables:

random variables whose ranges include intervals of the real numbers. We now give several examples of important discrete random variables.

Bernoulli random variables

These are the simplest possible random variables. Perform a random experiment with two possible outcomes: success or failure. Set $X = 1$ if the experiment is a success and $X = 0$ if the experiment is a failure. Such a 0–1 random variable is called a Bernoulli random variable. The usual notation is $P(X = 1) = p$ and $P(X = 0) = q = 1 - p$.

Example 2. Roll a fair die. We say that we have a success if we roll a 6. Thus, the probability of success is $P(X = 1) = 1/6$. We have $p = 1/6$ and $q = 5/6$.

Discrete uniform random variables

Example 3. Roll a fair die. Let X be the face shown. The distribution of X is given by the following table.

k	1	2	3	4	5	6
$P(X = k)$	1/6	1/6	1/6	1/6	1/6	1/6

Below we graph this distribution.

This is called a uniform random variable. *Uniform* refers to the fact that all possible values of X are equally likely.

Geometric random variables

Example 4. Roll a fair die until you get a 6. Let X be the number of rolls to get the first 6. The possible values of X are all strictly positive integers. Note that $X = 1$ if and only if the first roll is a 6. So $P(X = 1) = 1/6$. In order to have $X = 2$, the first roll must be anything but 6 and the second one must be 6. By independence of the different rolls we get $P(X = 2) = 5/6 \times 1/6$. More generally, in order to have $X = k$ the first $k - 1$ rolls cannot yield any 6 and the kth roll must be a 6. Thus,

$$P(X = k) = (5/6)^{k-1} \times 1/6 \text{ for all } k \geq 1.$$

Next, we graph this distribution.

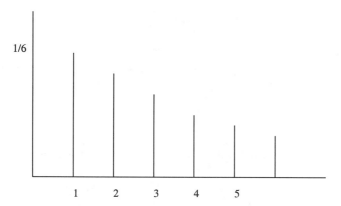

Such a random variable is called geometric. More generally, we have the following.

Geometric Random Variables

Consider a sequence of independent identical trials. Assume that each trial can result in a success or a failure. Each trial has a probability p of success and $q = 1 - p$ of failure. Let X be the number of trials before the first success. Then X is called a geometric random variable. The distribution of X is given by

$$P(X = k) = q^{k-1}p \text{ for all } k \geq 1.$$

Note that a geometric random variable may be arbitrarily large since the above probabilities are never 0. In order to check that the sum of these probabilities is 1 we need the following about geometric series from calculus:

Geometric Series

$$\sum_{k\geq 0} x^k = \frac{1}{1 - x} \text{ for all } x \in (-1, 1).$$

The sum of the distribution of X can thus be written

$$\sum_{k\geq 1} P(X = k) = \sum_{k\geq 1} q^{k-1}p = p\sum_{k\geq 0} q^k = \frac{p}{1 - q} = 1.$$

Example 5. Toss a fair coin until you get tails. What is the probability that exactly three tosses were necessary?

In this example we have $p = q = 1/2$. So

$$P(X = 3) = q^2 p = 1/8.$$

What is the probability that three or more tosses were necessary?

Note that the event '3 or more tosses are necessary' is the same as the event 'the first two tosses are heads'. Thus,

$$P(X \geq 3) = q^2 = 1/4.$$

Example 6. Consider X a geometric random variable. What is the probability that X is strictly larger than r?

The event '$X > r$' is the same as the event 'the first r trials are failures'. Thus,

$$P(X > r) = q^r.$$

Example 7. Let X be a geometric random variable. Given that $X > r$, what is the probability that $X > r + s$?

We want

$$P(X > r + s | X > r) = \frac{P(\{X > r + s\} \cap \{X > r\})}{P(X > r)}$$

where the equality comes from the definition of a conditional probability. Note that the intersection $\{X > r + s\} \cap \{X > r\}$ is simply $\{X > r + s\}$. Thus,

$$P(X > r + s | X > r) = \frac{P(X > r + s)}{P(X > r)}.$$

By Example 6, we know that $P(X > r) = q^r$. So

$$P(X > r + s | X > r) = \frac{q^{r+s}}{q^r} = q^s = P(X > s).$$

That is, given that we had r failures, the probability of getting an additional s failures is the same as getting s failures to start with. In this sense, the geometric distribution is said to have the *memoryless* property.

Example 8. Two players roll a die. If the die shows 6, then A wins; if the die shows 1 or 2, then B wins. The die is rolled until A or B wins. What is the probability that A wins?

Let T be the number of times the die is rolled. Note that the events $\{T = n\}$ are disjoint. Then

$$P(A) = \sum_{n \geq 1} P(A \cap \{T = n\}).$$

The event 'A wins in n rolls' is the same as the event 'the first $n-1$ rolls are draws and the nth roll is a 6.' Note that the probability that a roll results in a draw is 3/6. Then

$$P(A \cap \{T = n\}) = (1/2)^{n-1} \times 1/6.$$

Summing the geometric series we get

$$P(A) = \sum_{n \geq 1} (1/2)^{n-1} \times 1/6 = 1/3.$$

Note that the probability that A wins is

$$P(A) = 1/3 = \frac{1/6}{1/6 + 2/6}$$

where 1/6 is the probability of A winning in one roll and 2/6 is the probability of B winning in one roll.

Exercises

1. Toss three fair coins. Let X be the number of heads.
(a) Find the distribution of X.
(b) Compute $P(X \geq 2)$.

2. Roll two dice. Let X be the sum of the faces. Find the distribution of X.

3. Recall that there are 38 pockets in a roulette and that 18 are red. I bet on red until I win. Let X be the number of bets I make.
(a) What is the probability that X is 2 or more?
(b) What is the probability that X is exactly 2?

4. I roll four dice. I win if I get at least one 6. What is the probability of winning?

5. Roll two fair dice. Let X be the largest of the two faces. What is the distribution of X?

6. I draw two cards from a deck of 52. Let X be the number of aces I draw. Find the distribution of X.

7. How many times should I toss a fair coin in order to get tails at least once with probability 90%?

8. In a lottery there are 100 tickets numbered from 1 to 100. Let X be the number of the ticket drawn at random. What is the distribution of X?

9. I roll a die until I get a 6. Given that the first two rolls were not 6s, what is the probability that I need five rolls or more in order to get a 6?

10. A and B roll a die. A wins if the die shows a 6 and B wins if the die shows a 1. The die is rolled until someone wins.

(a) What is the probability that A wins?

(b) What is the probability that B wins?

(c) Let T be the number of times the die is rolled. Find the distribution of T.

11. Let X be a discrete random variable.

(a) Show that

$$P(X = k) = P(X > k - 1) - P(X > k).$$

(b) Assume that for all $k \geq 1$ we have $P(X > k) = q^k$. Use (a) to show that X is a geometric random variable.

2.2   Continuous Random Variables

We start with the following definition.

Continuous Random Variables

A continuous random variable is a function from a sample space Ω to an interval of the real numbers. The distribution of a continuous random variable X is determined by its *density* function f as follows. For all $a < b$,

$$P(a < X < b) = \int_a^b f(x)dx.$$

The function f is positive, continuous (except possibly at finitely many points) and

$$\int f(x)dx = 1,$$

where the integral is taken on the largest interval on which f is strictly positive.

The shaded area in the figure below represents the probability that the random variable is between 2 and 4.

Note that for a continuous random variable X the following probabilities are all equal.

$$P(a \leq X < b) = P(a \leq X \leq b) = P(a < X \leq b) = P(a < X < b).$$

This is so because integrals of the type $\int_a^a f(x)dx$ are always 0. In general, the above equalities do not hold for discrete random variables.

Example 1. Let X have density $f(x) = cx^2$ for x in $[-1, 1]$ and $f(x) = 0$ elsewhere. Find c.

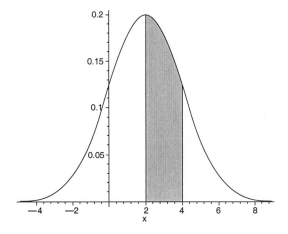

We must have

$$\int_{-1}^{1} cx^2 dx = 1.$$

After integrating we get

$$c(2/3) = 1$$

and therefore $c = 3/2$.

What is the probability that X is larger than 1/2?

$$P(X > 1/2) = \int_{1/2}^{1} f(x)dx = \int_{1/2}^{1} (3/2)x^2 dx = 7/16.$$

We next give two examples of important continuous random variables.

Continuous uniform random variables

Example 2. Let X be a random variable with density $f(x) = 1$ for x in [0,1] and $f(x) = 0$ elsewhere. Since the density of X is flat on [0,1], X is said to be uniform on [0,1]. Next we graph the density of X.

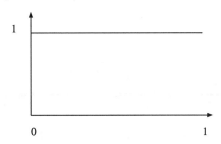

Note that

$$\int_{0}^{1} f(x)dx = \int_{0}^{1} dx = 1.$$

What is the probability of X being in the interval $(1/2, 3/4)$?

We have that

$$P(1/2 < X < 3/4) = \int_{1/2}^{3/4} f(x)dx = 1/4.$$

What is the probability that X is larger than $1/2$?

$$P(X > 1/2) = \int_{1/2}^{1} f(x)dx = 1/2.$$

More generally, we have the following.

Continuous uniform random variables

A continuous random variable X is uniform on the interval $[a, b]$ if the density of X is

$$f(x) = \frac{1}{b - a} \text{ for } x \in [a, b].$$

Note that the density of a uniform random variable is always a constant on some interval and that the constant must be picked so that the area under the density is 1.

Exponential random variables

Example 3. Let T be a random variable with density $f(t) = e^{-t}$ for $t \geq 0$. Below is the graph of f.

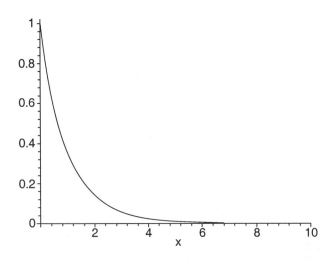

We first check that the area under the curve is 1.

$$\int_0^A e^{-t}dt = 1 - e^{-A}.$$

By letting A go to infinity we get that the improper integral converges and that

$$\int_0^\infty e^{-t}dt = 1.$$

What is the probability that T is larger than 1?

$$P(T > 1) = \int_1^\infty e^{-t}dt = e^{-1}.$$

What is the probability that T is less than 1?

$$P(T \le 1) = 1 - P(T > 1) = 1 - e^{-1}.$$

We next state the definition of an exponential random variable.

Exponential Random Variables

A random variable X with density $f(x) = ae^{-ax}$ for $x \ge 0$ is said to be an exponential random variable with parameter (or rate) $a > 0$.

Example 4. Let T be an exponential random variable with parameter a. What is the probability that T is larger than s?

$$P(T > s) = \int_s^\infty ae^{-at}dt = e^{-as}.$$

Example 5. Let T be an exponential random variable with parameter a. Given that T is larger than s, what is the probability that T is larger than $t + s$?

We want the conditional probability

$$P(T > t + s | T > s) = \frac{P(\{T > t + s\} \cap \{T > s\})}{P(T > s)}.$$

Note that the intersection of the events $T > t + s$ and $T > s$ is the event $T > t + s$. Thus,

$$P(T > t + s | T > s) = \frac{P(T > t + s)}{P(T > s)}.$$

By using the computation in Example 4 , we get

$$P(T > t + s | T > s) = \frac{e^{-a(t+s)}}{e^{-as}} = e^{-at} = P(T > t).$$

So exactly as for the geometric distribution of the preceding section, we say that the exponential distribution has the memoryless property.

Exercises

1. Let $f(x) = cx(1-x)$ for x in $[0,1]$ and $f(x) = 0$ elsewhere. Find c so that f is a density function.

2. Let the graph of the density f be a triangle for x in $[-1, 1]$. Find f.

3. Let X be the density of a uniform random variable on $[-2, 4]$. Find the density of X.

4. Let T be the waiting time for a bus. Assume that T has an exponential density with rate 3 per hour.

(a) What is the probability of waiting at least 20 minutes for the bus?

(b) Given that we have waited 20 minutes, what is the probability of waiting an additional 20 minutes for the bus?

(c) Under which conditions is the exponential model appropriate for this problem?

5. Let T be a waiting time for a bus. Assume that T has a uniform distribution on $[0,40]$.

(a) What is the probability of waiting at least 20 minutes for the bus?

(b) Given that we have waited 20 minutes, what is the probability of waiting an additional 10 minutes for the bus?

6. Let Y have a density $g(y) = cye^{-2y}$ for $y \geq 0$. Find c.

7. Let X have density $f(x) = xe^{-x}$ for $x \geq 0$. What is the probability that X is larger than 3?

8. Let T have density $g(t) = 4t^3$ for t in $[0,1]$.

(a) What is the probability that T is between 1/4 and 3/4?

(b) What is the probability that T is larger than 1/2?

9. (a) Show that for any random variable X we have

$$P(a < X < b) = P(X < b) - P(X \leq a).$$

(b) Assume that the random variable X is continuous and is such that $P(X > s) = e^{-2s}$. Use (a) to compute $P(a < X < b)$.

(c) Find the density of X.

2.3   Expectation

As we have seen in the preceding two sections, knowing the distribution of a random variable entails knowing a lot of information. For a discrete random variable X the distribution is given by the sequence $P(X = k)$ for every k in the range of X. For a continuous random variable X the distribution is given by the density function f. For many problems it is enough to have a rough idea of the distribution and one tries to summarize the distribution by using a few numbers. The most important of these numbers is the *expectation* or the average value of the distribution. We first deal with discrete random variables.

Expectation of a Discrete Random Variable

The expectation (or mean) of the discrete random variable X is denoted by $E(X)$ and is given by

$$E(X) = \sum_k k P(X = k)$$

where the sum is taken over all values in the range of X.

Note that the expectation of X is a measure of *location* of X. If a random variable may take infinitely many values, then the computation of its expectation involves an infinite series. The expectation is defined only if the infinite series converges (see Exercise 18).

Example 1. We perform an experiment with two possible outcomes: failure or success. If we have a success we set $X = 1$. If we have a failure we set $X = 0$. Let $P(X = 1) = p$. What is the expectation of this Bernoulli random variable?

$$E(X) = \sum_k k P(X = k) = 0 \times (1 - p) + 1 \times p = p.$$

Thus we can define,

Expectation of a Bernoulli Random Variable

Let X be a Bernoulli random variable with probability of success p. That is, X may take only values 0 and 1 and $P(X = 1) = p$. Then,

$$E(X) = p.$$

For instance, if we toss a fair coin and set $X = 1$ if we have heads and $X = 0$ if we get tails, then $E(X) = 1/2$. What is the meaning of the value 1/2, since X can only take values 0 and 1?

The Law of Large Numbers that we will now (loosely) describe gives a physical meaning to the notion of expectation.

Law of Large Numbers

We make n independent and identical random experiments. Each experiment has a random outcome with the same distribution as the random variable X. The Law of Large Numbers states that as n goes to infinity the average over the n outcomes approaches $E(X)$.

We now come back to Example 1. The Law of Large Numbers states that if we toss a coin many times, then the ratio of heads over the total number of tosses will approach 1/2. This gives a physical meaning to the expected value and also explains why this is a crucial notion.

Example 2. Roll a fair die. Let X be the face shown. We have $P(X = k) = 1/6$ for every $k = 1, 2, \ldots, 6$. Thus,

$$E(X) = \sum_k k P(X = k) = \sum_{k=1}^{6} k/6 = 7/2.$$

Example 3. The preceding example gave the expected value of a discrete uniform random variable in a particular case. We now treat the general case. Assume that X is a discrete uniform random variable on the set $\{1, 2, \ldots, n\}$. Thus, $P(X = k) = 1/n$ for $k = 1, 2, \ldots, n$. So

$$E(X) = \sum_{k=1}^{n} k P(X = k) = \frac{1}{n} \sum_{k=1}^{n} k.$$

Thus, we need to compute the sum of the first n integers. Let S_n be this sum and we write S_n in two different ways.

$$S_n = 1 + 2 + \cdots + (n-1) + n,$$

$$S_n = n + (n-1) + \cdots + 2 + 1.$$

We now add both equations. Note that the sum of the first terms on the right-hand side gives $n + 1$, likewise the sum of second terms gives $n + 1$ and so on. When we add the two equations we get n terms equal to $n + 1$ on the right-hand side. Thus,

$$2S_n = n(n + 1)$$

and we get

$$S_n = \frac{n(n + 1)}{2}.$$

Going back to the computation of the expected value we have

$$E(X) = \frac{n + 1}{2}.$$

Expectation of a Discrete Uniform Random Variable

Assume that X is a discrete uniform random variable on the set $\{1, 2, \ldots, n\}$.
Then
$$E(X) = \frac{n+1}{2}.$$

Note that if we let $n = 6$ we get the particular case of Example 1.

Example 4. We now deal with geometric random variables. Let X be the number of independent and identical trials to get the first success. We denote by p the probability that a given trial will be a success and $q = 1 - p$. The distribution of X is given by

$$P(X = k) = q^{k-1}p \text{ for all } k = 1, 2, \ldots.$$

Thus,

$$E(X) = \sum_{k=1}^{\infty} kq^{k-1}p = p\sum_{k=1}^{\infty} kq^{k-1}.$$

Recall that

$$\sum_{k=0}^{\infty} x^k = \frac{1}{1-x} \text{ for } x \in (-1, 1).$$

Recall also that power series are infinitely differentiable on their interval of convergence (except possibly at the boundary points). Thus, by taking derivatives on both sides of the preceding equality we get

$$\sum_{k=1}^{\infty} kx^{k-1} = \frac{1}{(1-x)^2} \text{ for } x \in (-1, 1).$$

We set $x = q$ and get for the expected value

$$E(X) = p\sum_{k=1}^{\infty} kq^{k-1} = p\frac{1}{(1-q)^2} = \frac{1}{p}.$$

Expectation of a Geometric Random Variable

Let X be the number of independent and identical trials to get the first success. We denote by p the probability that a given trial will be a success and $q = 1 - p$. Then,

$$E(X) = \frac{1}{p}.$$

Example 5. Roll a die until you get a 6. What is the expected number of rolls?

Let T be the number of rolls to get a 6. This is a geometric random variable with $p = 1/6$. Thus, $E(T) = 6$.

Continuous random variables

We start by defining the expected value for a continuous random variable.

Expectation of a Continuous Random Variable

Assume that X is a continuous random variable with density f. The expected value (or mean) of X is

$$\int xf(x)dx$$

where the integral is taken on the largest interval on which f is strictly positive.

Example 6. Assume that X is uniformly distributed on $[a, b]$. What is its expected value?

Using that the density of X is $f(x) = \frac{1}{b-a}$ for x in $[a, b]$, we get

$$E(X) = \int_a^b xf(x)dx = \frac{1}{b-a}(b^2/2 - a^2/2) = \frac{a+b}{2}.$$

Thus we can define

Expectation of a Continuous Uniform Random Variable

Assume that X is uniformly distributed on $[a, b]$. Then

$$E(X) = \frac{a+b}{2}.$$

Example 7. Assume T is exponentially distributed with rate a. What is its expected value?

We integrate by parts to get

$$E(T) = \int_0^\infty tf(t)dt = \int_0^\infty tae^{-at}dt = -te^{-at}\big]_0^\infty + \int_0^\infty e^{-at}dt = 1/a.$$

Expectation of an Exponential Random Variable

Assume T is exponentially distributed with rate a. Then,

$$E(T) = 1/a.$$

Other measures of location

To summarize the location of a distribution it is often a good idea to use more than one number. Besides the mean, there are two other important measures of location. The first one is the *median*.

Median of a Random Variable

A median m of a random variable X is a number m such that $P(X \leq m)$ and $P(X \geq m)$ are both at least $1/2$.

As we will show in the exercises a median gives less weight to the extreme values of the distribution than the mean.

Example 8. Roll a die. Let X be the face shown. Note that $P(X \geq 3) = 2/3$ and $P(X \leq 3) = 1/2$. So 3 is a median. Observe that 4 is also a median and actually any number in [3,4] is a median. Recall that the mean in this case is 3.5. This example shows that a discrete random variable may have several medians.

Unlike what may happen for discrete random variables, there is a unique median for continuous random variables. If the continuous variable X has density m, then the median of X is such that

$$\int_m^\infty f(x)dx = \int_{-\infty}^m f(x)dx = 1/2.$$

Example 9. Let T be an exponential random variable with rate 1. What is its median?

We solve the equation

$$P(T > m) = P(T \geq m) = \int_m^\infty e^{-t}dt = e^{-m} = 1/2.$$

Thus $m = \ln 2$. Note that $P(T < \ln 2) = 1 - P(T > \ln 2) = 1/2$. So $\ln 2$ is the unique median of this distribution.

Another measure of location, defined only for discrete random variables, is the *mode*.

> **Mode of a Discrete Random Variable**
>
> A mode M of a discrete random variable X is a number M such that $P(X = M)$ is maximum.

Example 10. For the uniform distribution on $\{1, 2, \ldots, 6\}$ there are six modes: $M = 1, 2, 3, 4, 5, 6$.

The addition rule

The following rule holds for any type (continuous or discrete) of random variables.

> **Addition Rule**
>
> Let X and Y be two random variables defined on the same sample space Ω. Then,
> $$E(X + Y) = E(X) + E(Y).$$
> More generally, if $X_1, X_2, \ldots, X_n$ are all defined on Ω we have
> $$E(X_1 + X_2 + \cdots + X_n) = E(X_1) + E(X_2) + \cdots + E(X_n).$$

As the next examples will show this is a very important rule. Its proof involves the joint distribution of several random variables. We will prove this formula when we see joint distributions in Section 8.3.

Example 11. I roll two dice. Let S be the sum of the two dice. What is the expected value of S?

Let X be the value of the first die and Y be the value of the second die. Then $S = X + Y$. According to the addition rule we have

$$E(S) = E(X) + E(Y).$$

But Example 1 tells us that $E(X) = E(Y) = 7/2$. Thus,

$$E(S) = 7.$$

We could have computed $E(S)$ by first computing the distribution of S and then averaging but this would have taken a lot longer.

Computing the expectation by breaking up the random variable

In many cases the distribution of a given random variable is too involved to be computed in a reasonable amount of time. In some of those cases it is possible to break up a random variable into a sum of Bernoulli random variables. By using the addition rule we then get the mean of the random variable with the specified distribution without computing that distribution. We next give such an example.

Example 12. Assume that three people enter independently an elevator that goes to five floors. What is the expected number of stops S that the elevator is going to make?

Instead of computing the distribution of S we break S into a sum of five Bernoulli random variables as follows. Let $X_1 = 1$ if at least one person goes to floor 1, otherwise we set $X_1 = 0$. Likewise let $X_2 = 1$ if at least one person goes to floor 2, otherwise we set $X_2 = 0$. We do the same for the five possible choices. We have

$$S = X_1 + X_2 + \cdots + X_5.$$

Note that $X_1 = 0$ if none of the three people pick floor 1. Thus,

$$P(X_1 = 0) = (4/5)^3.$$

The probability of success for X_1 is $p = P(X_1 = 1) = 1 - (4/5)^3$. All the X_i have the same Bernoulli distribution. By the addition rule we have

$$E(S) = 5p = 5(1 - (4/5)^3) = \frac{61}{25} = 2.44.$$

We now compute $E(S)$ by using the distribution of S. The random variable S may only take values 1, 2 and 3. In order to have $S = 1$, the second and third person need to pick the same floor as the first person. Thus,

$$P(S = 1) = (1/5)^2.$$

To have $S = 2$, there are two possibilities: either the second person picks the same floor as the first one and the third a different floor (the probability of that is $(1/5)(4/5)$) or the second person picks a different floor from the first one and the third one picks one of the two floors that have already been picked (the probability of that is $(4/5)(2/5)$). Thus,

$$P(S = 2) = (1/5)(4/5) + (4/5)(2/5).$$

Finally, $S = 3$ happens only if the three persons pick distinct floors:

$$P(S = 3) = (4/5)(3/5).$$

Thus,

$$E(S) = 1 \times \frac{1}{25} + 2 \times \frac{12}{25} + 3 \times \frac{12}{25} = \frac{61}{25}.$$

So even in this very simple case (S has only three values after all) it is better to compute the expected value of S by breaking S into a sum of 0–1 random variables rather than compute the distribution of S.

Example 13. Let B be the number of distinct birthdays in a class of 50 students. What is the $E(B)$?

The distribution of B is clearly fairly involved. We are going to break B into a sum of Bernoulli random variables. Set $X_1 = 1$ if at least one student was born January 1, otherwise set $X_1 = 0$. Set $X_2 = 1$ if at least one student was born January 2, otherwise set $X_2 = 0$. We define X_i as above for every one of the 365 days of the calendar. We claim that

$$B = X_1 + X_2 + \cdots + X_{365}.$$

This is so because the right-hand side counts all the days on which at least one student has his birthday. Moreover, the X_i are Bernoulli random variables. In order for $X_1 = 0$ we must have that none of the 50 students was born January 1. Thus,

$$P(X_1 = 0) = \left(\frac{364}{365}\right)^{50}.$$

The probability of success for X_1 is $p = 1 - (\frac{364}{365})^{50}$. We do the same for every X_i and they all have the same p (which is also the expected value of a Bernoulli random variable). By the addition rule we have

$$E(B) = E(X_1) + E(X_2) + \cdots + E(X_{365}) = 365p = 365\left(1 - \left(\frac{364}{365}\right)^{50}\right).$$

Numerically, we get

$$E(B) = 46.79.$$

From Example 2 in Section 1.4 we know that the probability of having at least two students born on the same day is 0.96. However from the value of $E(B)$ we see that more than two students born on the same day or more than one set of students born on the same day are not that likely, otherwise $E(B)$ would be lower.

Fair gambling

Example 14. We roll a die. You pay me b dollars if the die shows 5 or 6. I pay you one dollar otherwise. Clearly, the probabilities of winning are not the same for both players. Can we pick b so that this is a fair game?

Assume we play this game many times. By the Law of Large Numbers my average winnings will be close to my expected winnings. We will say that the game is fair if the expected winnings (of each player) are 0. So that in the long run my average winnings will approach 0.

In this particular case let W be my winnings in one bet. We have that $W = b$ with probability 1/3 and $W = -1$ with probability 2/3. Thus,

$$E(W) = b \times 1/3 + (-1) \times 2/3.$$

We want $E(W) = 0$. Solving for b we get $b = 2$. Since I am twice less likely to win than you are you should pay me twice as much when I win.

Expectation of a function of a random variable

As we will see in the next section it is often necessary to compute $E(X^2)$ for a random variable X. Unfortunately, this is *not* $E(X)^2$. We could compute the distribution of X^2 and use the distribution to compute the expected value. However, there is a quicker way to do things and it is contained in the following formula.

Expectation of a Function of X

Let X be a random variable and g be a real valued function. For instance, $g(x) = x^2$. Then if X is discrete

$$E(g(X)) = \sum_k g(k) P(X = k).$$

If X is continuous with density f, then

$$E(g(X)) = \int g(x) f(x) dx.$$

Example 15. Let X be a discrete random variable such that $P(X = -1) = 1/3$, $P(X = 0) = 1/2$ and $P(X = 2) = 1/6$. What is $E(X^2)$?
   We use the formula above with $g(x) = x^2$ to get

$$E(X^2) = (-1)^2 \times 1/3 + 0^2 \times 1/2 + (2)^2 \times 1/6 = 1.$$

Example 16. Let X be uniformly distributed on [0,1]. What is $E(X^3)$?
   This time we use the formula with $g(x) = x^3$. We get

$$E(X^3) = \int_0^1 x^3 f(x) dx = 1/4.$$

Another case of particular interest is when $g(x) = ax + b$. Assume that X is a discrete random variable. Then we use the formula above to get

$$
\begin{aligned}
E(aX + b) &= \sum_k (ak + b) P(X = k) \\
&= a \sum_k k P(X = k) + b \sum_k P(X = k) = a E(X) + b.
\end{aligned}
$$

The same formula may be derived for continuous random variables. We have the following for continuous and discrete random variables.

Expectation of a Linear Function of X

$$E(aX + b) = a E(X) + b.$$

Exercises

1. What is the expected value of a random variable uniformly distributed on $\{-1, 0, 3\}$.

2. Toss two fair coins. What is the expected number of heads?

3. The probability of finding someone in favor of a certain initiative is 0.01. We interview people at random until we find a person in favor of the initiative. What is the expected number of interviews?

4. Roll two dice. What is the expected value of the maximum of the two dice?

5. Let X be exponentially distributed with mean $1/2$. What is the density of X?

6. Let U be a random variable that is uniformly distributed on $[-1, 2]$.
   (a) Compute the mean of U.
   (b) What is the median of U?

7. Let X have the following density.

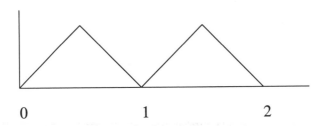

   (a) Find the expected value of X.
   (b) How good is $E(X)$ as a measure of location of X?

8. Let $f(x) = 3x^2$ for x in $[0,1]$. Let X be a random variable with density f.
   (a) What is $E(X)$?
   (b) What is the median of X?

9. Let X be a random variable such that $P(X = 0) = 1/5$ and $P(X = 4) = 4/5$. Find the mean, medians and modes.

10. Let T be exponentially distributed with rate a. Find the median of T as a function of a.

11. Roll four dice. What is the expected value of the sum?

12. There are three components in a circuit. Each one of them fails with probability p. The failure of one component may influence the other components in a way that is not well understood. What is the expected number of working components?

13. Let B be the number of distinct birthdays in a class of 200 students. What is the $E(B)$?

14. There are eight people in a bus and five bus stops ahead. What is the expected number of stops the bus will have to make for these eight people?

15. I roll four dice. If there is at least one 6 you pay me $1. If there are no 6s I pay you $1.
   (a) Is this a fair game?
   (b) How would you make it into a fair game?

16. Let X be uniform on $\{1, 2, \ldots, 6\}$. What is $E(X^2)$?

17. Let X be exponentially distributed with rate 1. What is $E(X^2)$?

18. In this problem we give an example of a discrete random variable for which the expectation is not defined.
   (a) Use the fact that

$$\sum_{k=1}^{\infty} \frac{1}{k^2} = \pi^2/6$$

to find c so that $P(X = k) = c/k^2$ is a probability distribution.
   (b) Show that the expectation of the random variable defined above does not exist.

19. This problem gives an example of a continuous random variable that has no expectation.
   (a) Show that $f(x) = \frac{2}{\pi(1+x^2)}$ for $x > 0$ is a density function.
   (b) Show that a random variable with the density above has no expectation.

2.4   Variance

We have seen in Section 2.3 that the expectation is a measure of location for a distribution. Next we are going to define a measure of dispersion: the variance. A small variance will mean that the distribution is concentrated around the mean and that the mean is a good measure of location. A large variance will mean that the distribution is dispersed and that no value is really typical for this distribution.

Variance of a Random Variable

Let X be a random variable with mean $E(X) = \mu$. The variance of X is denoted by $Var(X)$ and is defined as

$$Var(X) = E[(X - \mu)^2].$$

The following formula for the variance is useful for computational purposes.

$$Var(X) = E(X^2) - \mu^2.$$

Finally, the standard deviation of X is denoted by $SD(X)$ and is defined as

$$SD(X) = \sqrt{Var(X)}.$$

We now list some consequences of these definitions.

Consequences

C1. The variance of a random variable is *always* positive or 0. This is so because the variance is the expected value of the positive random variable $(X - \mu)^2$.

C2. The variance of a random variable X is 0 if and only if X is a constant. For a discrete random variable this can be seen from the formula

$$E[(X - \mu)^2] = \sum_k (k - \mu)^2 P(X = k).$$

If this sum is 0 it means that every term must be 0, since these are all positive terms. But the sum of the $P(X = k)$ is 1 so at least some of these terms are nonzero. It is easy to see that for exactly one k, $P(X = k)$ is not 0 and that corresponds to $k = \mu$. Thus, X is a constant equal to μ.

For a continuous random variable (that is a random variable whose density is strictly positive on some interval) one can show that the variance is always strictly positive.

C3. An easy consequence of the definition of variance is

Properties of Variance

For any random variable X and constants a and b,

$$Var(aX + b) = a^2 Var(X),$$

$$SD(aX + b) = |a| SD(X).$$

Observe that the translation by b has no effect on the variance of $aX + b$. Intuitively, this is clear since the variance measures the dispersion, not the location, of a random variable.

Example 1. We start with the Bernoulli distribution. Assume that X takes values 0 and 1. We denote $P(X = 1) = p$ and $P(X = 0) = 1 - p = q$. What is the variance of X?

We have that

$$E(X) = p.$$

We now compute

$$E(X^2) = 0^2 \times q + 1^2 \times p = p.$$

Thus,

$$Var(X) = E(X^2) - E(X)^2 = p - p^2 = pq.$$

Variance of a Bernoulli Random Variable

Assume that X takes values 0 and 1. We denote $P(X = 1) = p$ and $P(X = 0) = 1 - p = q$. Then,

$$Var(X) = pq.$$

Example 2. What is the variance of the discrete random variable uniformly distributed on $\{1, 2, 3, 4, 5, 6\}$?

We know that $E(X) = 7/2$.

We now compute

$$E(X^2) = 1^2 \times 1/6 + 2^2 \times 1/6 + \cdots + 6^2 \times 1/6 = \frac{91}{6}.$$

Thus,

$$Var(X) = E(X^2) - E(X)^2 = \frac{91}{6} - \frac{49}{4} = \frac{35}{12}.$$

So the standard deviation is approximately 1.7. It is large for a distribution on $\{1, \ldots, 6\}$. But this is not surprising since the extreme values have the same weight as the middle values for this distribution.

Example 3. We now turn to the variance of a geometric random variable. We make independent identical trials that have a probability p of success. Let T be the number of trials to get the first success. The random variable T has a geometric distribution and we know that

$$E(T) = 1/p.$$

As before we need to compute $E(T^2)$. In this case it is easier to compute $E(T(T-1))$ first. We need a new fact about geometric series. Recall that for every x in $(-1, 1)$ we have

$$\sum_{k=0}^{\infty} x^k = \frac{1}{1-x}.$$

Power series are infinitely differentiable on their interval of convergence. We take derivatives twice in the formula above to get:

$$\sum_{k=2}^{\infty} k(k-1)x^{k-2} = \frac{2}{(1-x)^3}. \tag{1}$$

Now we compute

$$E(T(T-1)) = \sum_{k=1}^{\infty} k(k-1)P(T=k) = \sum_{k=2}^{\infty} k(k-1)q^{k-1}p = pq\sum_{k=2}^{\infty} k(k-1)q^{k-2}.$$

We let $x = q$ in equation (1) to get

$$E(T(T-1)) = \frac{2pq}{(1-q)^3} = \frac{2q}{p^2}.$$

It follows that

$$E(T^2) = E(T(T-1)) + E(T) = \frac{2q}{p^2} + \frac{1}{p}.$$

Finally,

$$Var(T) = E(T^2) - E(T)^2 = \frac{2q}{p^2} + \frac{1}{p} - \frac{1}{p^2} = \frac{2q+p-1}{p^2}.$$

Note that $p + q = 1$, so $2q + p - 1 = q$. Hence,

$$Var(T) = \frac{q}{p^2}.$$

Variance of a Geometric Random Variable

Assume that we make independent identical trials that have a probability p of success. Let T be the number of trials to get the first success. Then,

$$Var(T) = \frac{q}{p^2}.$$

We now compute variances for a few continuous random variables.

Example 4. Assume that X is uniformly distributed on $[a, b]$. Then

$$E(X) = \frac{a + b}{2}.$$

We compute $E(X^2)$.

$$E(X^2) = \int_a^b x^2 f(x)dx = \frac{1}{b - a} \int_a^b x^2 dx = \frac{1}{3(b - a)}(b^3 - a^3) = \frac{b^2 + ab + a^2}{3}.$$

Thus,

$$Var(X) = E(X^2) - E(X)^2 = \frac{b^2 - 2ab + a^2}{12} = (b - a)^2/12.$$

Variance of a Continuous Uniform Random Variable

Assume that X is uniformly distributed on $[a, b]$. Then

$$Var(X) = (b - a)^2/12.$$

We now deal with exponential random variables.

Example 5. Assume that T is exponentially distributed with rate a. Then, $E(T) = 1/a$. We have

$$E(T^2) = \int_0^\infty t^2 f(t)dt = \int_0^\infty t^2 a e^{-at} dt.$$

We do an integration by parts to get

$$E(T^2) = -t^2 e^{-at}\big]_0^\infty + \int_0^\infty 2t e^{-at} dt = \frac{2}{a} \int_0^\infty t a e^{-at} dt = \frac{2}{a^2}$$

where we have used that $E(T) = 1/a$ to get the last equality. So

$$Var(T) = E(T^2) - E(T)^2 = 2/a^2 - 1/a^2 = 1/a^2.$$

That is, the mean and the standard deviation are equal for an exponential distribution. This shows that exponential distributions are rather dispersed.

Variance of an Exponential Random Variable

Assume that T is exponentially distributed with rate a. Then,

$$Var(T) = 1/a^2.$$

Example 6. Consider Y with the following density.

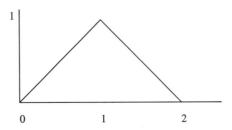

What is the $Var(Y)$?

The density of Y is $f(y) = y$ for y in $[0,1]$ and $f(y) = 2 - y$ for y in $[1,2]$. The mean of Y is 1 because of the symmetry of the density. We confirm this by computation.

$$E(Y) = \int_0^2 yf(y)dy = \int_0^1 y^2 dy + \int_1^2 y(2 - y)dy.$$

Thus,

$$E(Y) = y^3/3]_0^1 + y^2]_1^2 - y^3/3]_1^2 = 1.$$

We now deal with $E(Y^2)$.

$$E(Y^2) = \int_0^2 y^2 f(y)dy = \int_0^1 y^3 dy + \int_1^2 y^2(2 - y)dy,$$

So

$$E(Y^2) = y^4/4]_0^1 + 2y^3/3]_1^2 - y^4/4]_1^2 = 7/6,$$
$$Var(Y) = E(Y^2) - E(Y)^2 = 7/6 - 1 = 1/6.$$

Independent random variables

We will need to compute the variance of sums of random variables. This turns out to be a simple task only when the random variables in the sum are independent. We start by defining independence for random variables.

Independent Random Variables

Two discrete random variables X and Y are said to be independent if

$$P(\{X = x\} \cap \{Y = y\}) = P(X = x)P(Y = y) \text{ for ALL } x, y.$$

Two continuous random variables X and Y are said to be independent if for ALL real numbers $a < b, c < d$,

$$P(\{a < X < b\} \cap \{c < Y < d\}) = P(a < X < b)P(c < Y < d).$$

We now examine two examples.

Example 7. Roll two dice. Let X be the face shown by the first die and S be the sum of the two dice. Are X and S independent?

Intuitively it is clear that the answer should be no. It is enough to find one x and one y such that

$$P(\{X = x\} \cap \{S = y\}) \neq P(X = x)P(S = y)$$

in order to show that X and S are not independent. For instance, take $x = 1$ and $y = 12$. Clearly if one die shows 1, the sum cannot be 12. So $P(\{X = 1\} \cap \{S = 12\}) = 0$. However, $P(X = 1)$ and $P(S = 12)$ are strictly positive so $P(\{X = 1\} \cap \{S = 12\}) \neq P(X = 1)P(S = 12)$. X and S are not independent.

Example 8. Toss two fair coins. Set $X = 1$ if the first coin shows heads; set $X = 0$ otherwise. Set $Y = 1$ if the second coin shows heads; set $Y = 0$ otherwise. Are X and Y independent?

Our sample space is $\Omega = \{(H, H), (H, T), (T, H), (T, T)\}$. We need to examine the four possible outcomes for (X, Y). Note that the event $\{X = 0\} \cap \{Y = 0\}$ is the event $\{(T, T)\}$ and that it has probability 1/4. Note that $P(X = 0) = 2/4 = P(Y = 0)$. So the product rule holds for $x = 0$ and $y = 0$. We now examine $x = 0$ and $y = 1$. The event $\{X = 0\} \cap \{Y = 1\}$ is the event $\{(T, H)\}$. This has probability 1/4. Since $P(Y = 1) = 2/4$ the product rule holds in this case as well. The two remaining cases are symmetric to the cases we just examined. We may conclude that X and Y are independent.

Variance of a sum of random variables

If X and Y are independent it is easy to compute the variance of $X + Y$.

Variance of a Sum of Independent Random Variables

Assume that X and Y are two INDEPENDENT random variables defined on the same sample space Ω. Then,

$$Var(X + Y) = Var(X) + Var(Y).$$

More generally, if $X_1, X_2, \ldots, X_n$ are independent random variables, then

$$Var(X_1, X_2 + \cdots + X_n) = Var(X_1) + Var(X_2) + \cdots + Var(X_n).$$

Example 9. Roll two dice. Let S be the sum of the two dice. What is the variance of S?

Let X and Y be the faces shown by each die. From Example 2 we know that $Var(X) = Var(Y) = 35/12$. Since X and Y are independent we get that

$$Var(S) = Var(X + Y) = Var(X) + Var(Y) = 2 \times 35/12 = 35/6.$$

Example 10. Assume that X and Y are independent random variables with

$$Var(X) = 2 \qquad Var(Y) = 3.$$

What is the variance of $2X - 3Y$?

From the definition of independence it is easy to see that if X and Y are independent so are $2X$ and $-3Y$. Thus,

$$Var(2X - 3Y) = Var(2X) + Var(-3Y) = 4Var(X) + 9Var(Y) = 35.$$

Exercises

1. What is the variance of a random variable uniformly distributed on $\{-1, 0, 3\}$?

2. Let X be a random variable such that $P(X = 0) = 1/5$ and $P(X = 4) = 4/5$. Find the variance of X.

3. The probability of finding someone in favor of a certain initiative is 0.01. We interview people at random until we find a person in favor of the initiative. What is the standard deviation of the number of interviews?

4. Roll two dice.

(a) What is the variance of the maximum of the two dice?

(b) Compare the result of (a) to the variance of a single roll obtained in Example 2.

5. Let X have density $f(x) = x^2 e^{-x}/2$. What is the variance of X?

6. Let U be a random variable that is uniformly distributed on $[-1, 2]$. What is the variance of U?

7. Let X have the following density.

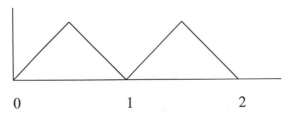

0	1	2

(a) Find the variance of X.

(b) Compare the result of (a) to the result in Example 6.

8. Let $f(x) = 3x^2$ for x in $[0,1]$. Let X be a random variable with density f. What is the variance of X?

9. Let X have variance 2. What is the variance of $-3X + 1$?

10. Let X be a measure in centimeters and let Y be the measure of the same object in inches. How are $SD(X)$ and $SD(Y)$ related?

11. Roll two dice successively. Let X be the face of the first die and Y the face of the second die.
  (a) Find $Var(X - Y)$.
  (b) Find $Var(|X - Y|)$.

12. A circuit has three components that work independently of each other with probability p_i for $i = 1, 2, 3$. Let S be the number of components that work. Find the variance of S.

2.5   Normal Random Variables

We start by giving the following definition.

Normal Random Variables

The continuous random variable X is said to be a normal random variable with mean μ and standard deviation σ if it has density

$$f(x) = \frac{1}{\sqrt{2\pi}\sigma} e^{-(x-\mu)^2/2\sigma^2}.$$

There are several things to be checked here: that f is a density, that $E(X) = \mu$ and that $Var(X) = \sigma^2$. Since these computations involve calculus only they will be left as exercises.

We graph below the densities of two normal random variables with $\mu = 2$. One has a standard deviation equal to 1 and the other one a standard deviation equal to 2. They both have the characteristic bell shaped form. However, one can see below how much more spread out the curve with $\sigma = 2$ is compared to the one with $\sigma = 1$.

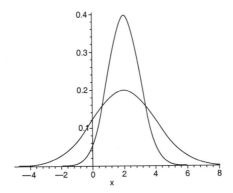

Standard Normal Random Variable

The continuous random variable Z is said to be a standard normal random variable if it has density

$$f(z) = \frac{1}{\sqrt{2\pi}} e^{-z^2/2}.$$

That is, Z is a normal random variable with mean 0 and standard deviation 1.

We graph below the density of a standard normal random variable.

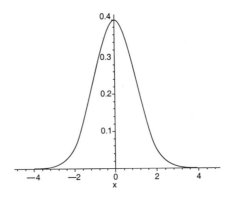

The notation Z will be reserved for standard normal random variables. In order to compute probabilities involving Z we will need to integrate its density. Unfortunately, there is no explicit formula for antiderivatives of $\frac{1}{\sqrt{2\pi}} e^{-z^2/2}$. We will need to rely on the numerical "Normal Table", provided as an appendix to the book for the function

$$\Phi(x) = P(0 < Z < x) = \int_0^x \frac{1}{\sqrt{2\pi}} e^{-z^2/2}.$$

Example 1. What is the probability that a standard normal random variable Z is larger than 1?

From the normal table we have that

$$P(Z > 1) = 1/2 - \Phi(1) = 1/2 - 0.34 = 0.16.$$

Example 2. What is the probability that a standard normal random variable Z is larger than -1?

By symmetry of the distribution of Z we have

$$P(Z > -1) = P(Z < 1) = 0.84.$$

Example 3. What is the value below which a standard normal random variable falls with probability 90%?

We want c such that

$$P(Z < c) = 1/2 + \Phi(c) = 0.9.$$

We see from the table that c is between 1.28 and 1.29. Since c is closer to 1.28, we take $c = 1.28$.

Example 4. What is the value below which a standard normal random variable falls with probability 20%?

This time we want c such that

$$P(Z < c) = 0.2.$$

Note that c is negative. By symmetry we have that

$$P(Z < c) = P(Z > -c) = 1/2 - \Phi(-c) = 0.2.$$

Thus,

$$\Phi(-c) = 0.3.$$

We read in the table that $-c$ is approximately 0.84. Thus, we have $c = -0.84$.

Example 5. What is the probability that a standard normal random variable Z is between -2 and 2?

$$P(-2 < Z < 2) = 2P(0 < Z < 2) = 2\Phi(2) \sim 0.95.$$

So there is only a 5% chance that a standard normal distribution is larger than 2 or smaller than -2.

One of the nice properties of the normal distributions is that they can easily be transformed into standard normal distributions as the property below shows.

Standardization
If X has normal distribution with mean μ and standard deviation σ, then the random variable $$\frac{X - \mu}{\sigma}$$ is a standard normal random variable.

What is remarkable here is not that $\frac{X-\mu}{\sigma}$ has mean 0 and standard deviation 1. This is true for any random variable that has a mean and a standard deviation, as will be shown below. What is remarkable is that after shifting and scaling a normal random variable we still get a normal random variable.

We now compute the expected value and standard deviation of $\frac{X-\mu}{\sigma}$.

$$E\left(\frac{X - \mu}{\sigma}\right) = \frac{1}{\sigma}(E(X) - \mu) = 0$$

where the last equality comes from the fact that $E(X) = \mu$. For the variance we have

$$Var\left(\frac{X - \mu}{\sigma}\right) = \frac{1}{\sigma^2}Var(X - \mu) = \frac{1}{\sigma^2}Var(X) = 1.$$

We now give a few examples of how to use the property above.

Example 6. Assume that heights of 6-year old are normally distributed with mean 100 cm and standard deviation 2 cm. What is the probability that a 6-year old taken at random is at least 105 cm tall?

Let X be the height of the child picked at random. We want $P(X > 105)$. We standardize X to get

$$P(X > 105) = P\left(\frac{X - 100}{2} > \frac{105 - 100}{2}\right) = P(Z > 2.5) \sim 0.01.$$

So there is only a 1% probability that a child taken at random will be at least 105 cm tall.

Example 7. What is the height above which 90% of the 6-year old are?

We want h such that $P(X > h)$. We standardize X again to get

$$P(X > h) = P\left(\frac{X - 100}{2} > \frac{h - 100}{2}\right) = P\left(Z > \frac{h - 100}{2}\right) = 0.9.$$

Note that $\frac{h-100}{2}$ must be negative. By symmetry of the distribution of Z we have that

$$P\left(Z > \frac{h - 100}{2}\right) = P\left(Z < \frac{-h + 100}{2}\right) = 0.9.$$

So according to the Normal Table we have

$$\frac{-h + 100}{2} = 1.28.$$

We solve for h and get that h is approximately 97.44 cm.

Example 8. Let X be normally distributed with mean μ and standard deviation σ. What is the probability that X is 2σ or more away from its mean?

We want

$$P(\{X > \mu + 2\sigma\} \cup \{X < \mu - 2\sigma\}) = P(X > \mu + 2\sigma) + P(X < \mu - 2\sigma)$$

where the last equality comes from the fact that the two events are disjoint. We standardize X to get

$$\begin{aligned}
P(\{X > \mu + 2\sigma\} \cup \{X < \mu - 2\sigma\}) &= P\left(\frac{X - \mu}{\sigma} > 2\right) + P\left(\frac{X - \mu}{\sigma} < -2\right) \\
&= P(Z > 2) + P(Z < -2) = 0.05.
\end{aligned}$$

Extreme observations

As we have just seen, the normal distribution is concentrated around its mean and it is unlikely that an observation taken at random is more than 2σ away from its mean (see Example 8). However, if we make several independent observations, what is the probability that the largest or the smallest of the observations is far away from the mean? We look next at a particular example.

Example 9. Assume that heights of 6-year olds are normally distributed with mean 100 cm and standard deviation 2cm. In a group of 25 children, what is the probability that the tallest of the group is at least 105 cm tall?

Let $X_1, \ldots, X_{25}$ be the heights of the 25 children in the group. We are interested in the probability that the maximum of these random variables will be at least 105. It is easier to deal with the complement of the preceding event. Note that the maximum of the 25 observations is less than 105 cm if and only if each one of the observations is less than 105 cm. Thus,

$$\begin{aligned}
P(\max(X_1, \ldots, X_{25}) < 105) &= P(\{X_1 < 105\} \cap \{X_2 < 105\} \cap \cdots \cap \{X_{25} < 105\}) \\
&= P(X_1 < 105)P(X_2 < 105) \ldots P(X_{25} < 105)
\end{aligned}$$

where the last equality comes from the independence of the X_i. According to Example 6, we have that $P(X_1 < 105)$ is $P(Z > 2.5) = 0.9876$ and this probability is the same for each X_i since they all have the same distribution. Thus,

$$P(\max(X_1, \ldots, X_{25}) < 105) = (0.9876)^{25} \sim 0.73.$$

That is, the probability that the tallest child in a group of 25 is at least 105 is 0.27. As the group increases, this probability increases as well. For a group of 50 this probability becomes about 0.5. For a group of 100 this probability becomes about 0.7.

The important conclusion of this example is the following. Extreme observations (especially if there are many of them) are likely to be far from a typical observation.

Exercises

1. Let Z be a standard normal random variable. Compute the following.
 (a) $P(Z > 1.52)$.
 (b) $P(Z > -1.15)$.
 (c) $P(-1 < Z < 2)$.
 (d) $P(-2 < Z < -1)$.

2. Let Z be a standard normal random variable. What is the value above which Z falls with 99% probability?

3. Assume that X is normally distributed with mean 3 and standard deviation 2.
 (a) $P(X > 3) =$?
 (b) $P(X > -1) =$?
 (c) $P(-1 < X < 3) =$?
 (d) $P(|X - 2| < 1) =$?

4. Assume that the diameter of a ball bearing is normally distributed with mean 1 cm and standard deviation 0.05 cm. A ball bearing is considered defective if its diameter is larger than 1.1 cm or smaller than 0.9 cm.
 (a) What is the proportion of defective ball bearings?
 (b) Find the diameter above which 99% of the diameters are.

5. Assume that X is normally distributed with mean 5 and standard deviation σ. Find σ so that $P(X > 4) = .95$.

6. Assume that the annual snow fall at some place is normally distributed with mean 20 inches and standard deviation eight inches.
 (a) What is the probability that the snow fall will be less than five inches in a given year?
 (b) What is the probability that the smallest annual snow fall in the next 20 years will be less than five inches?

7. Let Z be a standard normal random variable with density

$$f(z) = \frac{1}{\sqrt{2\pi}} e^{-z^2/2}.$$

In this exercise we will check that f is actually a density.
 (a) Change the variables from Cartesian to polar to show that

$$\int_{-\infty}^{+\infty} \int_{-\infty}^{+\infty} e^{-(x^2+y^2)/2} dx dy = \int_0^\infty \int_0^{2\pi} e^{-\rho^2/2} \rho d\rho d\theta.$$

(b) Show that the right-hand side of (a) is 2π.

(c) Show that the left-hand side of (a) is

$$\left(\int_{-\infty}^{+\infty} e^{-x^2/2} dx \right)^2.$$

(d) Conclude that f is a density.

8. Let Z be a standard normal random variable.

(a) Compute $E(Z)$.

(b) Compute $Var(Z)$.

9. Let

$$f(x) = \frac{1}{\sqrt{2\pi}\sigma} e^{-(x-\mu)^2/2}.$$

Show that f has inflection points at $\mu + \sigma$ and $\mu - \sigma$.

Review Exercises for Chapter 2

1. Three people toss one fair coin each. The winner is the one whose coin shows a face different from the two others. If the three coins show the same face, then there is a new round of tosses, until someone wins.

(a) What is the probability of exactly one round of tosses?

(b) What is the probability that at least three rounds of tosses are necessary?

2. A and B take turns rolling a die. A starts. The winner is the first one that rolls a 6. What is the probability that A wins?

3. Two people play the following game. They toss two fair coins. If the two coins land on heads then A wins. If one coin lands on heads, and the other on tails, then B wins. If the two coins land on tails, then the coins are tossed again until someone wins. What is the probability that B wins?

4. The probability of finding someone in favor of a certain initiative is 0.01. We interview people at random until we find a person in favor of the initiative. What is the probability that we need to conduct 50 or more interviews?

5. Draw five cards from a 52 cards deck.

(a) Explain why the probability that the second card is red is the same as the probability that the second card is black.

(b) What is the expected number of red cards among the five cards that have been drawn.

(c) What is the expected number of hearts in five cards dealt from a deck of 52 cards?

6. Assume that car battery lifetimes follow an exponential distribution with mean three years.

(a) What is the probability that a battery lasts 10 years or more?

(b) In a group of 10 batteries what is the probability that at least one will last 10 years or more?

(c) How many batteries do we need in order to have at least one last 10 years or more with probability 0.9?

7. Let X be a random variable with density $f(x) = ce^{-|x|}$.

(a) Find c.

(b) What is the $P(X > 1)$?

8. Let X have density $g(x) = c(x - 1)^2$ for x in [0,2].

(a) Find c.

(b) Find $E(X)$.

(c) Find $Var(X)$.

9. Let Y be a random variable with density $f(y) = c(-(y - 1)^2 + 2)$ for y in [0,2].

(a) Sketch the graphs of g in Exercise 8 and of f.

(b) Which random variable X or Y do you expect to have the highest variance?

(c) Confirm your prediction by doing a computation.

10. Roll two dice. I win one dollar if the sum is 7 or more. I lose b dollars if the sum is 6 or less. Find b so that this is a fair game.

11. Toss five fair coins.

(a) What is the expected number of heads?

(b) What is the variance of the number of heads?

12. Suppose atoms of a given kind have an exponential distributed lifetime with mean 30 years. What is the expected number of atoms still present after 30 years if we start with 10^{23} atoms?

13. Ball bearings are manufactured with diameters that are normally distributed with mean 1 cm and standard deviation 0.05 cm. Assume that 1,000 ball bearings are manufactured. What is the expected number of ball bearings whose diameter is at least 1.1 cm?

14. Assume that the random variable T is such that $E(T) = 1$ and $E(T(T-1)) = 2$. What is the standard deviation of T?

3

Binomial and Poisson Random Variables

3.1 Counting Principles

Before stating the fundamental principle of counting we give an example.

Example 1. Assume that a restaurant offers five different specials and for each one of them you can pick either a salad or a soup. How many choices do you have?

In this simple example we can just enumerate all the possibilities. Number the specials from 1 to 5 and let S denote the salad and O denote the soup. There are 10 possibilities:

$$(1, S) \quad (2, S) \quad (3, S) \quad (4, S) \quad (5, S)$$
$$(1, O) \quad (2, O) \quad (3, O) \quad (4, O) \quad (5, O)$$

This is so because we have two selections to make, one with two choices and the other one with five choices. Thus, in all there are $2 \times 5 = 10$ choices.

The Multiplication Rule

If we have r successive selections with n_k choices at the kth step, for $k = 1 \ldots r$, then in all we have $n_1 \times n_2 \times \cdots \times n_r$ possibilities.

Example 2. Consider an answer sheet with five categories for age, two categories for sex, three categories for education. How many possible answer sheets are there?

In this example we have $r = 3$, $n_1 = 5$, $n_2 = 2$ and $n_3 = 3$. Thus, in all there are $5 \times 2 \times 3 = 30$ possibilities.

Example 3. In a true/false test there are 10 questions. How many different ways can this test be answered?

This time we have 10 successive selections to be made and for each selection we have two choices. In all there are $2 \times 2 \times \cdots \times 2 = 2^{10}$ possibilities.

Example 4. How many arrival orders are there for three runners?

We call the three runners A, B and C. A has three possible arrival positions. Once the arrival of A is fixed then B has only two possible arrival positions. Once the arrivals of A and B are fixed there is only one possible arrival position for C. Thus, we may use the multiplication rule to get that in all there are $3 \times 2 \times 1 = 6$ possibilities.

The preceding example illustrates a particularly important consequence of the multiplication rule.

Permutations

For any positive integer n define n factorial as

$$n! = n \times (n-1) \times (n-2) \times \cdots \times 1 \text{ for } n \geq 1$$

and

$$0! = 1.$$

A particular labeling of n distinct objects is called a permutation of these n objects. The number of possible permutations of n objects is $n!$.

Note that in Example 4 we are counting the number of permutations of three runners. The number of permutations is $3! = 6$.

Example 5. How many ways are there to put seven different books on a shelf?

Again we need to count the number of permutations of seven distinct objects. We get $7! = 5040$ possibilities.

Note that the factorials can be computed inductively by using the formula

$$n! = n \times (n-1)!.$$

Factorials grow very fast (see Exercise 10).

In many situations we want to pick a set of (non-ordered) k objects among n objects where $k \leq n$. How many ways are there to do that?

Let $\binom{n}{k}$ (it is read 'n choose k') be the number of ways to pick a subset of k objects among n objects. For the first object we pick we have n choices, for the second one we have $n-1$ choices, for the third one $n-2$ choices and so on. For the kth object we have $(n-k+1)$ choices. So according to the multiplication rule we have $n \times (n-1) \times (n-2) \times \cdots \times (n-k+1)$ ways to pick an *ordered* set of k objects. We know that a set of k objects has $k!$ permutations. That is, for every set of k objects there are $k!$ ways to order it. Thus, we have a rule:

The number of ways to pick an ordered set of k objects $= k! \binom{n}{k}$.

So

$$n \times (n-1) \times (n-2) \times \cdots \times (n-k+1) = k! \binom{n}{k}.$$

Observe that

$$n \times (n-1) \times (n-2) \times \cdots \times (n-k+1) = \frac{n!}{(n-k)!}.$$

Thus,

$$\binom{n}{k} = \frac{n!}{k!(n-k)!}.$$

Ordered and Non-ordered Sets

The number of ways to pick an ordered set of k elements out of n elements is

$$n \times (n-1) \times (n-2) \times \cdots \times (n-k+1).$$

A particular way to pick a non-ordered set of k elements out of n is called a combination. The number of combinations of k elements out of n is given by

$$\binom{n}{k} = \frac{n!}{k!(n-k)!}.$$

Example 6. Three awards will be given to three distinct students in a group of 10 students. How many ways are there to give these three awards?

We want to know how many subsets of three students can be picked out of a set of 10 students. This is exactly

$$\binom{10}{3} = \frac{10!}{3!7!} = \frac{10 \times 9 \times 8}{3 \times 2} = 120.$$

Example 7. In a contest, 10 students will be ranked and the top three will get gold, silver and bronze medals, respectively. How many ways are there to give these three medals?

This is different from Example 6 because the order of the three students picked is important. There are 10 possible choices for the gold medal, there are nine choices for the silver medal and there are eight choices for the bronze. So according to the multiplication rule there are $10 \times 9 \times 8$ ways to give these medals. That is 720 ways. Note that this is six (that is, 3!) times more ways than in Example 6.

Example 8. In a business meeting seven people shake hands. How many handshakes are there in all?

There are as many handshakes as there are sets of two people among seven. So the number is

$$\binom{7}{2} = \frac{7!}{2!5!} = 21.$$

Example 9. How many distinct strings of letters can be made out of the word CARE?

Every permutation of these four distinct letters will give a distinct string of letters. Thus, there are $4! = 24$ distinct strings of letters.

Example 10. How many distinct strings of letters can be made out of the word PEPPER?

There are 6! possible permutations of these six letters. However, there are only four distinct letters in this word. For instance, if we permute the Ps only (there are 3! such permutations) we get the same string of letters. If we permute the Es only (there are 2! such permutations) we also get the same string. Thus, there are

$$\frac{6!}{2!3!} = 60$$

distinct strings of letters.

Example 11. How many distinct strings of 0s and 1s can we make with three 1s and two 0s?

This is exactly the same problem as Example 10. Note that there are 5! permutations but since there are three 1s and two 0s the total number of distinct strings is:

$$\frac{5!}{3!2!} = \binom{5}{2} = 10.$$

Example 12. You are dealt five cards from a 52 cards deck. What is the probability of getting a full house (three of a kind and a pair of another kind)?

We first observe that there are $\binom{52}{5}$ equally likely hands. Next we use the multiplication rule. There are 13×12 ways to pick two distinct kinds (one for the pair, another one for the triplet). Once we have picked the pair kind there are $\binom{4}{2}$ choices to make a pair. For the triplet there are $\binom{4}{3}$ choices. So there are

$$13 \times 12 \times \binom{4}{2} \times \binom{4}{3}$$

ways to pick a full house. Assuming that all hands are equally likely we get that the probability of a full house is

$$\frac{13 \times 12 \times \binom{4}{2} \times \binom{4}{3}}{\binom{52}{5}} \sim 0.001 \, .$$

Example 13. You are dealt five cards from a 52 cards deck. What is the probability of getting three of a kind?

There are $\binom{13}{1}$ ways to pick the kind for the triplet. Once the kind of the triplet is picked there are $\binom{4}{3}$ ways to pick three cards to make a triplet. There are $\binom{12}{2}$ ways to pick the two remaining kinds. Note that this is *not* 12×11. This is so because the two remaining cards are exchangeable: a Queen and a King is the same as a King and a Queen for the two remaining cards. Once the kind of each remaining card has been picked, then there are $\binom{4}{1}$ to pick a card for each kind. Thus, the number of ways to pick three of a kind is

$$\binom{13}{1}\binom{4}{3}\binom{12}{2}\binom{4}{1}\binom{4}{1}.$$

By dividing the formula above by $\binom{52}{5}$ we get a probability of 0.02.

Properties of the binomial coefficients

The $\binom{n}{k}$ are also called binomial coefficients because of their role in the binomial theorem which we will see below. We start by listing a few useful properties of these coefficients.

P1. Recall that $0!=1$ so

$$\binom{n}{0} = 1 \text{ for every integer } n \geq 0.$$

P2. For all integers $n \geq 1$ and $k \geq 1$ we have that

$$\binom{n}{k} = \binom{n-1}{k-1} + \binom{n-1}{k}.$$

In order to see the preceding identity, fix a particular element out of the n elements we have and call it O. We have two possible types of subsets of k elements. The ones that contain O and the ones that do not contain O. There are $\binom{n-1}{k-1}$ subsets of k elements that contain O. This is so because if we pick O, then we need to pick $k - 1$ elements out of $n - 1$. There are $\binom{n-1}{k}$ subsets of k elements that do not contain O. By adding the two preceding binomial coefficients we get all the subsets of k elements out of n. This proves P2.

P3. For all integers $n \geq 0$ and $k \geq 0$ we have that

$$\binom{n}{k} = \binom{n}{n-k}.$$

Each time we pick k out of n elements, we do not pick $n - k$ out of n elements. So there are as many subsets with k elements as there are with $n - k$ elements. This proves P3.

P4. Pascal's triangle. This is a convenient way to compute the binomial coefficients by using the preceding properties.

$$
\begin{array}{c|cccccc}
\mathbf{k} & 0 & 1 & 2 & 3 & 4 & 5 \\
\hline
\mathbf{n} & & & & & & \\
0 & 1 & & & & & \\
1 & 1 & 1 & & & & \\
2 & 1 & 2 & 1 & & & \\
3 & 1 & 3 & 3 & 1 & & \\
4 & 1 & 4 & 6 & 4 & 1 & \\
5 & 1 & 5 & 10 & 10 & 5 & 1
\end{array}
$$

One reads $\binom{n}{k}$ at the intersection of row n and column k. For instance, $\binom{4}{2} = 6$. The triangle is constructed by using Property P2. For instance,

$$
\binom{4}{2} = \binom{3}{1} + \binom{3}{2}.
$$

That is, we get 6 by adding the 3 immediately above and the 3 above and to the left. Note that Pascal's triangle is symmetric and that is a consequence of P3.

We now turn to the binomial theorem.

Binomial Theorem

For any integer $n \geq 0$ and any real numbers a and b,

$$
(a+b)^n = \sum_{k=0}^{n} \binom{n}{k} a^k b^{n-k}.
$$

We will see why the theorem holds on a particular example. Take $n = 4$; then

$$
(a+b)^4 = (a+b) \times (a+b) \times (a+b) \times (a+b).
$$

All the terms in the expansion come from these four products. So all the terms must have degree 4. That is all the terms are of the type $a^i b^j$ where $i + j = 4$. To get a^4 we must pick a in each one of the four terms in the product, and there is only one way to do that. In the final expansion there is only one a^4. To get $a^3 b$ we need to pick as from three of the four terms in the product, and there are $\binom{4}{3} = 4$ ways to do that. In the final expansion there are 4 $a^3 b$. To get $a^2 b^2$ we need to pick 2 as and there are $\binom{4}{2} = 6$ ways to do that. Using the symmetry property P3 we get

$$
(a+b)^4 = a^4 + 4a^3 b + 6a^2 b^2 + 4ab^3 + b^4.
$$

Exercises

1. Someone has three pairs of shoes, two pairs of pants and four shirts. In how many ways can he get dressed?

2. A test is composed of 12 questions. Each question can be true, false or blank. How many ways are there to answer this test?

3. In how many ways can seven persons stand in line?

4. How many five cards hands are there out of a deck of 52?

5. A child has five balls. Two are red and three are blue. How many ways are there to place the balls in a straight line?

6. License plates have three letters and four numbers. How many different license plates can be made?

7. In a class of 21, in how many ways can a professor give out three As?

8. In a class of 21, in how many ways can a professor give out three As and three Bs?

9. Assume that eight horses are running and that three will win.
(a) How many ways are there to pick the unordered three winners?
(b) How many ways are there to pick the ordered three winners?

10. According to Stirling's formula we have that
$$n! \sim \sqrt{2\pi} n^{n+1/2} e^{-n}.$$

That is, the ratio of the two sides tends to 1 as n goes to infinity. Use Stirling's formula to approximate 10!, 20! and 50!. How good are these approximations?

11. Use Pascal's triangle to compute $\binom{10}{k}$ for $k = 0 \ldots 10$.

12. You are dealt five cards from a 52 cards deck. What is the probability that
(a) you get exactly one pair?
(b) you get two pairs?
(c) you get a straight flush (five consecutive cards of the same suit)?
(d) a flush (five of the same suit but not a straight flush)?
(e) a straight (five consecutive cards but not a straight flush)?

13. (a) Show that
$$\sum_{k=0}^{n} \binom{n}{k} = 2^n.$$
(b) Use (a) to show that a set of n elements has 2^n subsets.

14. Compute
$$\sum_{k=0}^{n} \binom{n}{k} (-1)^k.$$

15. Expand $(x + y)^7$.

3.2 Binomial Random Variables

Recall that a Bernoulli random variable X is a random variable with two possible outcomes, usually denoted by 0 and 1. Think of 0 as being a failure and 1 as being a success. Assume that $P(X = 1) = p$ and $P(X = 0) = 1 - p$. Consider n independent and identically distributed Bernoulli random variables $X_1, X_2 \ldots X_n$ and let B be the number of successes among these n experiments. In other words, we have that

$$B = X_1 + X_2 + \cdots + X_n.$$

The random variable B is said to have a binomial distribution with parameters n and p.

We are now going to derive the distribution of B. One of the ways B may be equal to k is if the first k Bernoulli random variables are successes and the last $n - k$ are failures. This happens with probability $p^k(1 - p)^{n-k}$. However, there are as many ways for $B = k$ as there are ways to distribute k 1s and $n - k$ 0s among n places. This is the same problem as the one we solved in Example 11 in 3.1. We want the number of distinct strings of 0s and 1s that have length n and k 1s. There are $n!$ ways to arrange n distinct objects in n places. In the present case there are only two types of objects: 0s and 1s. For each distribution of 1s and 0s there are $k!(n - k)!$ permutations that correspond to the same distribution. Thus, there are $\frac{n!}{k!(n-k)!}$ ways to distribute k 1s and $n - k$ 0s in n places. Thus,

$$P(B = k) = \binom{n}{k} p^k(1 - p)^{n-k}.$$

We now summarize these facts about the binomial distribution in the box below.

Binomial Random Variables

Consider n independent and identically distributed Bernoulli random variables $X_1, X_2 \ldots X_n$. Let $P(\text{success in the } i\text{th trial}) = P(X_i = 1) = p$, for $i = 1 \ldots n$. Let B be the number of successes among these n experiments. That is,

$$B = X_1 + X_2 + \cdots + X_n.$$

The random variable B is said to have a binomial distribution with parameters n and p. It follows that,

$$P(B = k) = \binom{n}{k} p^k(1 - p)^{n-k}.$$

Note that for a binomial B with parameters n and p the formula simplifies for the extreme values

$$P(B = 0) = (1 - p)^n \text{ and } P(B = n) = p^n$$

and that by the Binomial Theorem,

$$\sum_{k=0}^{n} P(B = k) = \sum_{k=0}^{n} \binom{n}{k} p^k (1 - p)^{n-k} = (p + 1 - p)^n = 1.$$

Example 1. Roll a fair die five times. What is the probability of getting exactly two 6s?

In this case we are doing $n = 5$ identical experiments. The probability of success is $p = 1/6$ and B is the number of 6s (or successes) we get in five trials. Thus,

$$P(B = 2) = \binom{5}{2} (1/6)^2 (5/6)^3 = 10 \frac{5^3}{6^5} \sim 0.16.$$

Example 2. What is the probability of getting at least one 6 in five rolls?

We want the probability of $\{B \geq 1\}$. It is quicker to compute the probability of the complement of $\{B \geq 1\}$ which is $\{B = 0\}$.

$$P(B = 0) = (1 - p)^n = (5/6)^5 \sim 0.4.$$

Thus, the probability of getting at least one 6 in five rolls is approximately 0.6.

Example 3. Assume that births of boys and girls are equally likely. What is the probability that a family with three children have three girls?

This time we have $n = 3$ trials and each has a probability of success (having a girl) equal to $p = 1/2$. We want

$$P(B = 3) = (1/2)^3 = 1/8.$$

Example 4. Consider four families, each with three children. What is the probability that exactly one family has three girls?

We have $n = 4$ trials and a trial is a success if the corresponding family has exactly three girls. According to Example 3 the probability of success is 1/8. Thus,

$$P(B = 1) = \binom{4}{1} (1/8)^1 (7/8)^3 \sim 0.33.$$

Binomial coefficients grow very fast. Next we give an algorithm that allows the computation of a binomial distribution while avoiding the computation of the binomial coefficients.

Computational Formula for the Binomial Distribution

Let B be a binomial random variable with parameters n and p. Then

$$P(B = 0) = (1 - p)^n$$

and

$$P(B = k) = \frac{p}{1 - p}\frac{n - k + 1}{k}P(B = k - 1) \text{ for } k = 1, 2 \ldots, n.$$

We derive the preceding formula. Let $k \geq 1$,

$$
\begin{aligned}
P(B = k) &= \binom{n}{k}p^k(1 - p)^{n-k} = \frac{n!}{k!(n - k)!}p^k(1 - p)^{n-k} \\
&= \frac{p}{1 - p}\frac{n - k + 1}{k}\frac{n!}{(k - 1)!(n - k + 1)!}p^{k-1}(1 - p)^{n-k+1} \\
&= \frac{p}{1 - p}\frac{n - k + 1}{k}P(B = k - 1).
\end{aligned}
$$

We now apply the preceding formula to an example.

Example 5. Find the distribution of a binomial random variable with $n = 8$ and $p = 0.2$.

We have that

$$P(B = 0) = (1 - p)^n = (0.8)^8 \sim 0.17.$$

We use the recursion for $k \geq 1$,

$$P(B = k) = \frac{p}{1 - p}\frac{n - k + 1}{k}P(B = k - 1) = \frac{1}{4}\frac{8 - k + 1}{k}P(B = k - 1).$$

For instance,

$$P(B = 1) = \frac{1}{4}8P(B = 0) \sim 0.34.$$

We summarize the distribution in the table below.

k	0	1	2	3	4	5	6	7	8
$P(N = k)$	0.17	0.34	0.29	0.15	0.05	0.01	0.001	0	0

We now turn to the mean and variance of the binomial distribution.

Mean and Variance of a Binomial Distribution

Assume that B is a binomial random variable with parameters n and p. Then

$$E(B) = np$$

and

$$Var(B) = np(1 - p) = npq.$$

Recall that a binomial random variable B is a sum of independent identically distributed Bernoulli random variables. That is,

$$B = X_1 + X_2 + \cdots + X_n.$$

Thus,

$$E(B) = E(X_1) + E(X_2) + \cdots + E(X_n)$$

and using that $E(X_i) = p$ for $i = 1 \ldots n$ we get

$$E(B) = np.$$

Recall that $Var(X_i) = p(1 - p) = pq$ and that the variance of the sum of *independent* random variables is the sum of the variances. Thus,

$$Var(B) = Var(X_1) + \cdots + Var(X_n) = npq.$$

Example 6. Roll a die 30 times what is the expected number of 5s?

The number of 5s is a binomial random variable with parameters $n = 30$ and $p = 1/6$. So the expected number of 5s is $np = 5$.

Example 7. Assume that 100 components have exponential lifetimes with mean one year. Assume that the components fail independently one of the other. What is the expected number of components that have not failed after two years?

Let B be the number of components that have not failed after two years. We may write

$$B = X_1 + \cdots + X_{100}$$

where $X_i = 1$ if the ith component has not failed after two years and $X_i = 0$ otherwise, for $i = 1 \ldots 100$. The X_i are independent identically distributed Bernoulli random variables with probability of success

$$p = \int_2^\infty e^{-t} dt = e^{-2}.$$

So B is a binomial random variable with parameters 100 and p and the expected value of B is

$$E(B) = np = 100e^{-2} \sim 13.53.$$

Another measure of location is the mode. Recall that a mode M (not necessarily unique) of a discrete random variable B is such that $P(B = M)$ is the maximum of all the $P(X = k)$.

Mode of a Binomial Random Variable

Let B be a binomial random variable with parameters n and p. If $np + p$ is not an integer, then there is a unique mode M which is the greatest integer less than $np + p$. If $np + p$ is an integer, then there are two modes $np + p$ and $np + p - 1$.

See Exercise 10 for a proof of the formula above.

Example 8. Roll a die 30 times. What is the most likely number of 6s we will get?

The number of 6s is a binomial random variable with parameters $n = 30$ and $p = 1/6$. We first examine $np + p = 5 + 1/6$. This is not an integer, therefore we have a unique mode: the largest integer below $5 + 1/6$, that is 5. The most likely number of 6s is five.

Note that if we roll the die 35 times, then $np + p = 6$ and we have two modes $np + p = 6$ and $np + p - 1 = 5$. So if we roll the die 35 times there are two most likely numbers of 6s: five and six.

Normal approximation to the binomial distribution

As n becomes large the computation of something like $P(B \geq a)$ may involve computing many binomial probabilities. The most important technique to deal with this problem is the normal approximation.

Normal Approximation

Let B be a binomial distribution with parameters n and p. As n increases the distribution of $\frac{B - np}{\sqrt{npq}}$ approaches the distribution of a standard normal random variable Z in the sense that for any $a \leq b$ we have

$$P(a \leq B \leq b) \sim P\left(\frac{a - np - 1/2}{\sqrt{npq}} \leq Z \leq \frac{b - np + 1/2}{\sqrt{npq}}\right) \text{ as } n \to \infty.$$

We are using a continuous random variable Z to approximate a discrete random variable B. This is why we enlarge the interval by 1/2 on both sides. This is especially important if $a = b$ or if $\sqrt{npq}$ is small.

As the figures below illustrate, the larger n is the closer a binomial histogram is to a normal curve. Another fact that can be seen below is that the convergence towards the normal curve is faster when p is closer to 1/2.

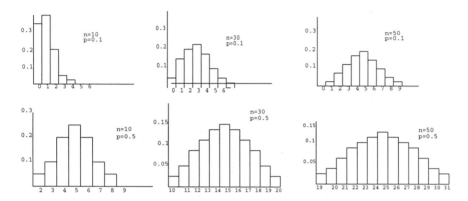

Example 9. Roll a fair die 36 times; what is the probability that we get exactly six 6s?

Let B be the number of 6s we get in 36 rolls. Then B is a binomial distribution with parameters 36 and 1/6. We first compute the exact probability.

$$P(B = 6) = \binom{36}{6} (1/6)^6 (5/6)^{30} \sim 0.176$$

We now use the normal approximation. Note that $np = 6$ and $npq = 5$.

$$P(B = 6) \sim P\left(\frac{6 - 6 - 1/2}{\sqrt{5}} \leq Z \leq \frac{6 - 6 + 1/2}{\sqrt{5}}\right) \sim 0.174.$$

So even in this example with n not so large and p not close to 1/2, the approximation is good.

Example 10. A hotel has accepted 210 reservations but it has only 200 rooms. It is assumed that guests will actually show up independently of each other with probability 0.9. What is the probability that the hotel will not have enough rooms?

Let B be the number of guests that will actually show up. This is a binomial random variable with parameters 210 and 0.9. The mean number of guests showing up is $np = 189$ and the variance is $npq = 18.9$. The normal approximation yields

$$P(201 \leq B) \sim P\left(\frac{201 - 189 - 1/2}{\sqrt{18.9}} \leq Z\right) = P(2.64 \leq Z) \sim 0.004.$$

It is rather unlikely that not enough rooms will be available.

Example 11. Assume that a fair coin is tossed 10,000 times. Let B be the number of heads. What is the probability of getting exactly 5,000 heads?

We use the normal approximation to get

$$P(B = 5000) \sim P\left(\frac{5000 - np - 1/2}{\sqrt{npq}} \leq Z \leq \frac{5000 - np + 1/2}{\sqrt{npq}}\right).$$

The mean is $np = 5,000$ and the standard deviation $\sqrt{npq} = 50$. Thus,

$$P(B = 5000) \sim P\left(-0.01 \leq Z \leq 0.01\right) \sim 0.008.$$

So the probability of getting exactly 5,000 heads is rather slim: less than 1%.

Note that $np + p = 5,000 + 1/2$ and so the most likely number of heads is 5,000. However, there are so many possible values that any fixed number of heads is rather unlikely.

Example 12. Assume that a fair coin is tossed 10,000 times. Let B be the number of heads. Find a so that B is between $E(B) - a$ and $E(B) + a$ with probability 0.99.

The expected value for B is $np = 10,000 \times 1/2 = 5,000$. We want a so that

$$P(E(B) - a \leq B \leq E(B) + a) \sim P\left(\frac{-a - 1/2}{\sqrt{npq}} \leq Z \leq \frac{a + 1/2}{\sqrt{npq}}\right) = 0.99.$$

Using the normal table we get

$$\frac{a + 1/2}{\sqrt{npq}} = 2.57.$$

Thus,

$$a = 2.57\sqrt{npq} - 1/2.$$

In this particular case we get $a = 128$. So with 99% confidence, the number of heads will be in the interval $[5,000 - 128; 5,000 + 128]$, which is rather narrow considering that we are performing 10,000 tosses. The important lesson of this example is that the number of successes of a binomial with parameters n and p is in the interval $(np - (2.57\sqrt{npq} - 1/2), np + (2.57\sqrt{npq} - 1/2))$ with probability 0.99 when n is large. In particular, typical deviations from the mean are of order $\sqrt{n}$.

The negative binomial

Example 13. Roll a fair die. What is the probability that the second 6 appears at the 10th roll?

Note that the event $A = \{$ the second 6 appears at the 10th roll $\}$ is the intersection of the two events $B = \{$ there is exactly one 6 in the first nine rolls $\}$ and $C = \{$ the 10th roll is a 6$\}$. Moreover, B and C are independent since B depends on the first nine rolls and C depends on the 10th roll. Note that the number of 6s in nine rolls is a binomial with parameters 9 and 1/6. Moreover, $P(C) = 1/6$. Thus,

$$P(A) = P(B)P(C) = \binom{9}{1}(1/6)(5/6)^8(1/6) = 9\frac{5^8}{6^9} \sim 0.06.$$

More generally, we have the following.

Negative Binomial

Assume that we perform identical and independent trials, each trial having a probability of success equal to p. Let r be an integer larger than or equal to 1. Let B_r be the number of trials until the rth success. Then B_r is called a negative binomial random variable with parameters r and p. Moreover,

$$P(B_r = k) = \binom{k-1}{r-1}p^{r-1}(1-p)^{k-r}p = \binom{k-1}{r-1}p^r(1-p)^{k-r} \text{ for } k = r, r+1, \ldots.$$

Note that in the case $r = 1$, B_1 is the number of trials until the first success. This is exactly a geometric random variable and the formula above simplifies to

$$P(B_1 = k) = p(1-p)^{k-1} \text{ for } k = 1, 2, \ldots.$$

Example 14. What is the probability that the fifth child of a couple is their second girl?

This is a negative binomial with $r = 2$ and $p = 1/2$. Thus,

$$P(B_2 = 5) = \binom{4}{1}(1/2)(1/2)^3(1/2) = \frac{4}{2^5} = 1/8.$$

We now turn to the mean and variance of a negative binomial random variable.

Mean and Variance of a Negative Binomial

Let B_r be a negative binomial random variable with parameters r and p. Then,

$$E(B_r) = \frac{r}{p}$$

$$Var(B_r) = \frac{r(1-p)}{p}.$$

We derive the formulas above by breaking B_r into a sum of simpler random variables. Let G_1 be the number of trials until the first success; G_1 is a geometric random variable with parameter p. Let G_2 be the number of trials from the first success until the second success. The random variable G_2 is also geometric and G_1 and G_2 are independent. More generally, we define G_i to be the number of trials between the $(i-1)$th success and the ith success for $i = 1, 2, \ldots, r$. It is easy to see that

$$B_r = G_1 + G_2 + \cdots + G_r.$$

All the G_i are independent and identically distributed according to a geometric distribution. Recall that $E(G_1) = 1/p$. Thus,

$$E(B_r) = rE(G_1) = \frac{r}{p}.$$

By using the independence of the G_i and the fact that $Var(G_1) = (1-p)/p$ we get

$$Var(B_r) = rVar(G_1) = \frac{r(1-p)}{p}.$$

Example 15. Roll a fair die. How many rolls are expected to get the third 6?

The number of trials to get the third 6 is a negative binomial with parameters $r = 3$ and $p = 1/6$. So

$$E(B_3) = \frac{3}{1/6} = 18.$$

Exercises

1. Toss a fair coin four times.
   (a) What is the probability of getting at least one heads?
   (b) What is the probability of getting at least three heads?
   (c) What is the probability of getting exactly two heads?

2. Roll two fair dice five times.
   (a) What is the probability of getting at least one sum equal to 7?
   (b) What is the probability of getting at least two sums larger than or equal to 7?

3. Toss a fair coin seven times. Let B be the number of heads.
   (a) Draw the histogram of the distribution of B.
   (b) What is the mean of B?
   (c) What is the mode of B?

4. Toss a fair coin 11 times. What is the most likely number of heads?

5. Given that there were five heads in 12 tosses of a fair coin.
   (a) What is the probability that the first toss was heads?
   (b) What is the probability that the last two tosses were heads?

(c) What is the probability that at least two of the first five tosses were heads?

6. Assume that 100 components have normal lifetimes with mean one year and standard deviation six months. Assume that the components fail independently one of the other.

(a) What is the probability that at least two components have not failed after two years?

(b) What is the expected number of components that have not failed after two years?

7. Assume that 500 invitations have been sent out for a given event. Assume that each person shows up independently of the others with probability 0.6.

(a) What is the probability that 250 or fewer people show up?

(b) Find b so that the number of people that show up is b or larger with probability 0.9.

8. Roll a fair die 360 times.

(a) What is the probability to get exactly sixty 1s?

(b) Find a so that the number of 1s is in the interval $[60 - a, 60 + a]$ with probability 95%.

9. Toss a fair coin 100 times.

(a) What is the probability of getting exactly 50 heads?

(b) Assume that 25 probability students toss a fair coin 100 times each. What is the probability that at least one student gets exactly 50 heads?

10. In this exercise we derive the formulas for the mode. Let B be a binomial with parameters n and p.

(a) By using the computational formula for the binomial distribution, show that $P(B = k - 1) \leq P(B = k)$ if and only if $k \leq np + p$.

(b) By definition of the mode M we must have simultaneously $P(B = M - 1) \leq P(B = M)$ and $P(B = M + 1) \leq P(B = M)$. Use (a) to show that

$$np + p - 1 \leq M \leq np + p.$$

(c) Show that there is only one integer M solution of the double inequality in (b) when $np + p$ is not an integer.

(d) Show that there are two solutions to the double inequality in (b) when $np + p$ is an integer.

11. In 1975, in Columbus Ohio there were 12 cases of childhood leukemia. The expected number is six per year (Morbidity and mortality weekly report, July 25 1997, p. 671–674). Assume that there are 200,000 children under 15 in that area and that each one has the same probability 3×10^{-5} of being hit by leukemia in a given year.

(a) Use the computational formula for the binomial distribution to compute the probability of having 12 or more cases of leukemia in a given year.

(b) Assume that there are 200 regions in the United States with the same number of children and the same probability for each child to be struck by leukemia. What is the probability that at least one region will get 12 cases or more?

(c) Considering (a) and (b), would you attribute the cluster in Columbus to chance alone?

12. Toss a fair coin.

(a) What is the probability that the third head occurs at the 8th toss?

(b) What is the expected number of tosses to get the 10th head?

13. Items are examined sequentially at a manufacturing plant. The probability that an item is defective is 0.05.

(a) What is the probability that the first 20 items examined are not defective?

(b) What is the expected number of examined items until we get the fifth defective?

14. What is the probability that the fifth child of a couple is their third girl?

3.3   Poisson Random Variables

We start with the definition.

Poisson Random Variables

The random variable N is said to have a Poisson distribution with mean λ if

$$P(N = k) = e^{-\lambda}\frac{\lambda^k}{k!} \text{ for } k = 0, 1, \ldots.$$

Later in this section we will show that a random variable with the distribution above has indeed mean λ. Typically the Poisson distribution appears when we count the number of occurrences of events that have small probabilities and are independent.

Example 1. Consider a fire station that serves a given neighborhood. Each resident has a small probability of needing help on a given day and most of the time people need help independently one of the other. The number of calls a fire station gets on a given day may be approximated by a Poisson random variable with mean λ. The parameter λ may be taken to be the observed average. Assume that $\lambda = 6$. What is the probability that a fire station gets two or more calls in a given day?

$$P(N \geq 2) = 1 - P(N = 0) - P(N = 1) = 1 - e^{-\lambda} - \lambda e^{-\lambda} = 1 - 7e^{-6} \sim 0.98.$$

Example 2. Assume that a book has an average of one misprint every 10 pages. What is the probability that a given page has no misprint?

Consider all the words in a given page; we may assume that each one of them has a small probability of being misprinted. We may also assume that each word is misprinted independently of the other words. With these assumptions the Poisson distribution is adequate. The mean number of misprints per page is $\lambda = 0.1$. Thus,

$$P(N = 0) = e^{-\lambda} = e^{-0.1} \sim 0.9.$$

The next property shows that a binomial distribution with parameters n and p may be approximated by a Poisson distribution with mean $\lambda = np$.

Poisson Approximation of the Binomial

Let B be a binomial random variable with parameters n and p. Let N be a Poisson random variable with mean $\lambda = np$; then for every $k \geq 0$,

$$P(B = k) \sim P(N = k) \text{ for small } p.$$

The smaller p is the better the approximation above is; for more details see Hodges and Le Cam, *Annals of Mathematical Statistics* (1960), p. 737–740. Thanks to the Poisson approximation we replace a two parameters distribution by a one parameter distribution and we avoid the computation of binomial coefficients. Note that if B is a binomial random variable with parameters n and p, then

$$P(B = 0) = (1 - p)^n.$$

Recall from calculus that

$$\lim_{p \to 0} \frac{\ln(1 - p)}{-p} = 1.$$

Therefore,

$$P(B = 0) = (1 - p)^n = \exp(n \ln(1 - p)) \sim \exp(-np) = P(N = 0)$$

where the approximation holds for p small enough. In order to prove that a binomial with small p may be approximated by a Poisson, we need to show that for every $k \geq 0$ it is true that $P(B = k) \sim P(N = k)$. For a proof see the reference above.

Example 3. During a recent meteor shower it was estimated that the probability of a given satellite to be hit by a meteor is 1/1000. Assuming that there are 500 satellites around the Earth and that they get hit independently one of the other, what is the probability that no satellite will be hit?

Let B be the number of satellites hit. Under these assumptions B has a binomial distribution with parameters 500 and 1/1000. Thus

$$P(B = 0) = (1 - 1/1000)^{500} \sim 0.6064.$$

We now use the Poisson approximation with $np = 1/2$.

$$P(N = 0) = 1 - e^{-\lambda} = e^{-1/2} \sim 0.6065.$$

One can see that the approximation is excellent in this case.
  What is the probability that two or more satellites are hit?
  This time we want

$$P(N \geq 2) = 1 - P(N = 0) - P(N = 1) = 1 - e^{-\lambda} - \lambda e^{-\lambda} = 1 - \frac{3}{2}e^{-1/2} \sim 0.09.$$

  The next example will use the following algorithm to compute the distribution of a Poisson random variable.

Computational Formula for the Poisson Distribution

Let N be a Poisson random variable with mean λ; then its distribution may be computed inductively by using the algorithm

$$P(N = 0) = e^{-\lambda},$$

$$P(N = k) = \frac{\lambda}{k}P(N = k - 1) \text{ for all } k \geq 1.$$

The formula above is easy to derive. Assume $k \geq 1$, then

$$P(N = k) = e^{-\lambda}\frac{\lambda^k}{k!} = e^{-\lambda}\frac{\lambda}{k}\frac{\lambda^{k-1}}{(k-1)!} = \frac{\lambda}{k}P(N = k - 1).$$

Example 4. Assume that a hospital serves 100,000 people and that each person may get hit by a certain disease with probability 3×10^{-5} per year, independently one of the other. What is the probability that the hospital will see six or more cases of the disease in a given year?
  Under the assumptions, the number of cases of the disease follows a binomial distribution with parameters $n = 100,000$ and $p = 3 \times 10^{-5}$. We use the Poisson approximation with mean $\lambda = np = 3$. We want

$$P(N \geq 6) = 1 - \sum_{k=0}^{5} P(N = k).$$

We use the computational formula to get

k	0	1	2	3	4	5
$P(N = k)$	0.05	0.15	0.22	0.22	0.17	0.1

Thus,

$$P(N \geq 6) \sim 1 - 0.91 = 0.09.$$

In many situations we may need more involved models than the simple binomial in Example 4. For instance, in the case of cancer the probability of getting hit increases significantly with age. So a more realistic model should separate people in age classes. The total number of cancer cases is then a sum of binomial random variables with different ps. This is not a binomial random variable. However, the next result shows that we may still use the Poisson approximation when all the ps are small.

Poisson Approximation of a Sum of Binomial Random Variables

Let B_i, for $i = 1 \ldots r$, be independent binomial random variables with parameters n_i and p_i. Let

$$\lambda = n_1 p_1 + \cdots + n_r p_r$$

and N be a Poisson random variable with mean λ. Then for every $k \geq 0$,

$$P(B_1 + B_2 + \cdots + B_r = k) \sim P(N = k) \text{ when all the } p_i \text{s are small.}$$

Example 5. Assume that a hospital serves 100,000 people that are in three different class ages. Assume that an individual in class i has a probability p_i (independently of all the other individuals) of getting a certain disease. Class 1 has $n_1 = 50,000$ individuals and $p_1 = 2 \times 10^{-5}$, class 2 has $n_2 = 30,000$ individuals and $p_2 = 5 \times 10^{-5}$ and class 3 has $n_3 = 20,000$ individuals and $p_3 = 10^{-4}$. What is the probability that on a given year this hospital sees three or more cases of the disease?

For each class i the number of cases B_i follows a binomial distribution with parameters n_i and p_i. We are interested in the event $B_1 + B_2 + B_3 \geq 3$. Since the B_i are independent and the p_is are small, we may use the Poisson approximation. Let

$$\lambda = n_1 p_1 + n_2 p_2 + n_3 p_3 = 4.5$$

and let N be a Poisson random variable with mean λ. We then have

$$
\begin{aligned}
P(B_1 + B_2 + B_3 \geq 3) \sim P(N \geq 3) &= 1 - (P(N = 0) + P(N = 1) + P(N = 2)) \\
&= 1 - e^{-\lambda} - \lambda e^{-\lambda} - \lambda^2 e^{-\lambda}/2 \sim 0.83.
\end{aligned}
$$

We now turn to a property that shows that the Poisson distribution is bound to appear in many situations.

Poisson Scatter Theorem

Consider a finite interval I that gets random hits (the interval may represent a time interval and the hits may represent incoming telephone calls). Assume the following two hypotheses: (1) a given point of I may get hit at most once and (2) for any $n \geq 1$, if we divide I in n equal parts, then each of the n subintervals gets hit with the same probability and independently of the other $n - 1$ subintervals. Under (1) and (2) there is $\lambda > 0$ such that the total number of hits on I follows a Poisson distribution with mean λ. Let L be the length of I; then any subinterval of I with length ℓ has a Poisson distribution with mean $\lambda \ell / L$.

For a proof of this theorem, see *Probability* by J. Pitman (Springer-Verlag, 1997).

Example 6. Consider a telephone exchange on a Monday from 2:00 to 3:00 PM. Assume that there is an average of 120 calls during this time period. What is the probability of getting at least four calls in a three minutes interval?

It may be reasonable to assume that hypotheses (1) and (2) hold (the only question about this is whether each subinterval of time is equally likely to get calls). Then according to the Poisson Scatter Theorem, the number of calls during a three minute interval follows a Poisson distribution with mean $120 \times 3/60 = 6$.

$$
\begin{aligned}
P(N \geq 4) &= 1 - (P(N = 0) + P(N = 1) + P(N = 2) + P(N = 3)) \\
&= 1 - e^{-6} - 6e^{-6} - \frac{6^2}{2}e^{-6} - \frac{6^3}{3!}e^{-6} \sim 0.85.
\end{aligned}
$$

The Poisson Scatter Theorem holds in any dimension. For instance, it may be used to count the number of stars that appear on a photographic plate or the number of raisins in a cookie. In the first case we replace length by area and in the second one we replace length by volume.

Example 7. Assume that rain drops are hitting a square with side 10 inches. Assume that the average is 30 drops per minute. What is the probability that a subsquare with side 2 inches does not get hit in a given minute?

Again it seems reasonable to assume that hypotheses (1) and (2) hold. The number of rain drops in the subsquare follows a Poisson distribution with mean $30 \times 2^2/10^2 = 1.2$. Thus,

$$
P(N = 0) = e^{-1.2} \sim 0.3.
$$

Example 8. Assume that a given document has on average two misprints per page. Given that there are no misprints in the first half of a page, what is the probability that there will be two or more misprints in the second half of this page?

It is reasonable to assume that hypotheses (1) and (2) hold and therefore the number of misprints in the two half pages are independent. Let A be the event 'there are no misprints in the first half of the page' and let B be the event 'there are at least two misprints in the second half of the page'. Let N be the number of misprints in the second half page. According to the Poisson Scatter Theorem, N follows a Poisson distribution with mean $2 \times 1/2 = 1$. Thus, we have

$$P(B|A) = P(B) = P(N \geq 2) = 1 - (P(N = 0) + P(N = 1)) = 1 - 2e^{-1}.$$

For some of the computations below it will be useful to recall the following from calculus.

> **Taylor Series for the Exponential Function**
>
> $$e^x = \sum_{k=0}^{\infty} \frac{x^k}{k!} \text{ for every } x.$$

In particular we see that if N is a Poisson random variable with mean λ, then

$$\sum_{k=0}^{\infty} P(N = k) = \sum_{k=0}^{\infty} e^{-\lambda} \frac{\lambda^k}{k!} = e^{-\lambda} e^{\lambda} = 1.$$

We are now going to compute the mean and variance of a Poisson random variable N with mean λ.

$$E(N) = \sum_{k=0}^{\infty} k P(N = k) = \sum_{k=1}^{\infty} k e^{-\lambda} \frac{\lambda^k}{k!} = e^{-\lambda} \lambda \sum_{k=1}^{\infty} \frac{\lambda^{k-1}}{(k-1)!}.$$

By shifting the summation index we get

$$\sum_{k=1}^{\infty} \frac{\lambda^{k-1}}{(k-1)!} = \sum_{k=0}^{\infty} \frac{\lambda^k}{k!} = e^{\lambda}.$$

Thus,

$$E(N) = e^{-\lambda} \lambda \sum_{k=1}^{\infty} \frac{\lambda^{k-1}}{(k-1)!} = e^{-\lambda} \lambda e^{\lambda} = \lambda.$$

We now turn to the computation of the variance of N. It turns out that it is easier to compute $E(N(N-1))$ than $E(N^2)$.

$$E(N(N-1)) = \sum_{k=0}^{\infty} k(k-1) P(N = k) = \sum_{k=2}^{\infty} k(k-1) e^{-\lambda} \frac{\lambda^k}{k!} = e^{-\lambda} \lambda^2 \sum_{k=2}^{\infty} \frac{\lambda^{k-2}}{(k-2)!}.$$

We shift again the summation index to get

$$E(N(N-1)) = e^{-\lambda} \lambda^2 \sum_{k=2}^{\infty} \frac{\lambda^{k-2}}{(k-2)!} = e^{-\lambda} \lambda^2 e^{\lambda} = \lambda^2.$$

So $E(N(N-1)) = \lambda^2$ and therefore

$$E(N^2) = E(N(N-1) + E(N) = \lambda^2 + \lambda.$$

By definition of the variance we have

$$Var(N) = E(N^2) - E(N)^2 = \lambda.$$

We now summarize the computations above.

Mean and Variance of a Poisson Random Variable

Let N be a Poisson random variable with mean λ; then

$$E(N) = Var(N) = \lambda.$$

Example 9. We have seen that the distribution of binomial random variable B with parameters n and p may be approximated by a Poisson distribution with mean $\lambda = np$ if p is small. We also know that B may be approximated by a normal distribution if n is large. This implies that if n is large and p small, then a Poisson random variable with mean λ may be approximated by a normal distribution. In fact, the larger λ the better the approximation. In this example, we will compute $P(N = 5)$ exactly and by using a normal approximation for a Poisson random variable N with mean 7. The exact computation is

$$P(N = 5) = e^{-7}\frac{7^5}{5!} \sim 0.13.$$

We now use the normal approximation.

$$\begin{aligned}
P(N = 5) &= P(4.5 \le N \le 5.5) \\
&= P\left(\frac{4.5 - E(N)}{SD(N)} \le \frac{N - E(N)}{SD(N)} \le \frac{5.5 - E(N)}{SD(N)}\right).
\end{aligned}$$

We now approximate the distribution of $\frac{N-E(N)}{SD(N)}$ by the distribution of a standard normal distribution Z. Thus,

$$P(N = 5) \sim P\left(\frac{4.5 - 7}{\sqrt{7}}\right) \le Z \le \frac{5.5 - 7}{\sqrt{7}}\right) = P(-0.94 \le Z \le -0.57) \sim 0.11.$$

Mode of a Poisson Random Variable

Let N be a Poisson random variable with mean λ. If λ is an integer, then N has two modes: λ and $\lambda - 1$. If λ is not an integer, then N has a unique mode: the largest integer smaller than λ.

For a proof see Exercise 10 below.

Exercises

1. Assume that books from a certain publisher have an average of one misprint every 20 pages.

    (a) What is the probability that a given page has two or more misprints?

    (b) What is the probability that a book of 200 pages has at least one page with two or more misprints?

2. Suppose that cranberry muffins have an average of six cranberries.

    (a) What is the probability that half a muffin has at least four cranberries?

    (b) What is the probability that half a muffin has 2 or less cranberries?

    (c) Given that the first half of my muffin had two cranberries or less, what is the probability that the second half has four or more cranberries?

3. Assume that you bet 200 times on 7 at the roulette (there are 38 possible slots). What is the probability that you win at least three times?

4. (a) Use the computational formula to compute $P(N = k)$ for $k = 0, 1, \ldots, 10$ for a Poisson random variable with mean $\lambda = 5$.

    (b) What are the modes of the distribution in (a)?

5. Assume that 1000 individuals are screened for a condition that affects 1% of the general population. What is the probability that exactly 10 individuals have the condition?

6. Assume that an elementary school has 500 children.

    (a) What is the probability that at least one child was born on April 15?

    (b) What is the probability that at least three children were born on April 15?

7. The number of incoming phone calls at a telephone exchange is modelled using a Poisson distribution with mean $\lambda = 2$ per minute.

    (a) What is the probability of having five or fewer calls in a three minutes interval?

    (b) Given that there were seven calls in the first three minutes, what is the probability that there were no calls during the first minute?

    (c) Show that given that there were n calls during the first t minutes, the number of calls during the first $s < t$ minutes follows a binomial with parameters s/t and n.

8. Suppose that the probability of a genetic disorder is 0.05 for men and 0.01 for women. Assume that 50 men and 100 women are screened.

    (a) Compute the exact probability that exactly two individuals among the 150 that have been screened have the disorder.

    (b) Use the Poisson approximation for a sum of binomial random variables to compute the approximate probability of the event in (a).

9. Assume that 1% of men under 20 experience hair loss and that 10% of men over 30 experience hair loss. A sample of 20 men under 20 and 30 men over 30 are examined. What is the probability that 4 or more men experience hair loss?

10. In this problem we are going to find a formula for the mode of a Poisson distribution.

(a) Use that

$$P(N = k) = \frac{\lambda}{k} P(N = k - 1) \text{ for } k \geq 1$$

to show that $P(N = k) \geq P(N = k - 1)$, if and only if $\lambda \geq k$.

(b) Show that $P(N = k) \geq P(N = k + 1)$, if and only if $\lambda \leq k + 1$.

(c) Show that the mode M of N must satisfy the double inequality

$$\lambda - 1 \leq M \leq \lambda.$$

(d) Show that if λ is an integer, then there are two modes λ and $\lambda - 1$. Show that if λ is not an integer, then there is a unique mode which is the largest integer smaller than λ.

11. Let N be a Poisson random variable with mean 10.

(a) What is the exact probability that $N = 10$?

(b) Use the normal approximation of Example 9 to compute the probability in (a).

Review Exercises for Chapter 3

1. How many distinct strings of letters can we get from the word TOUGH?

2. How many distinct strings of letters can we get from the word PROBABILITY?

3. Roll three dice.

(a) What is the probability of getting a sum equal to 10?

(b) What is the probability of getting a sum equal to 9?

4. Assume you toss a coin 100 times and you get 32 heads. Do you think this is a fair coin? (*Hint*: assume it is a fair coin and compute the probability of getting 32 or fewer heads).

5. Roll a pair of dice 10 times.

(a) What is the probability to get at least once a pair of 6s?

(b) What is the probability of getting twice a pair of 6s?

(c) What is the probability of getting the first pair of 6s at the 10th roll?

6. Roll a die.

(a) What is the probability of getting the first 6 at or before the fifth roll?

(b) What is the probability of getting the third 6 at the 10th roll?

(c) What is the expected number of rolls to get the fifth 6?

(d) Given that the second 6 occurred at the 10th roll, what is the probability that the first 6 occurred at the fifth roll?

7. A gambler bets repeatedly $1 on red at the roulette (there are 18 red slots and 38 slots in all). He wins $1 if red comes up loses $1 otherwise. What is the probability that he will be ahead
  (a) after 100 bets?
  (b) after 1000 bets?

8. Assume that each passenger shows up independently of the others with probability 0.95. How many tickets should the airline sell for a flight on an airplane with 200 seats so that, with probability 0.99, each passenger that shows up gets a seat on the flight?

9. A company has three factories A, B and C. A has manufactured 1000 items, B has manufactured 1,500 items and C has manufactured 2,000 items. Assume that the probability that an item is defective is 0.003 for A, 0.002 for B and 0.001 for C. What is the probability that the total number of defective items is seven or larger?

10. Assume that lamp bulbs have exponential life times with mean two years. What is the probability that in a box of ten:
  (a) exactly two will last at least two years?
  (b) none will last more than one year?
  (c) What is the expected number of lamp bulbs that will last at least 2 years?

11. Assuming that boys and girls are equally likely, how many children should a couple plan to have in order to have at least one boy and one girl with probability 0.99?

12. In average there is one defect per 100 meters of magnetic tape.
  (a) What is the probability that 150 m of tape have no defect?
  (b) Given that the first 100 m of tape had no defect, what is the probability that the whole 150 m has no defect?
  (c) Given that the first 100 m of tape had at least one defect, what is the probability that the whole 150 m has exactly 2 defects?

13. Assume you bet $1 100 times on 7 (there are 38 equally likely slots). If 7 comes up you win $35, otherwise you lose your $1.
  (a) What are your expected winnings?
  (b) What is the probability that you are ahead after 100 bets?
  (c) What is the probability that you have lost $100?

14. Assume you bet $1 100 times on red (there are 38 equally likely slots and 18 are red). If red comes up you win $1, otherwise you lose your $1.
  (a) What are your expected winnings?
  (b) What is the probability that you are ahead after 100 bets?
  (c) What is the probability that you have lost $100?

15. Assume that 49 students each toss a fair coin 100 times.
  (a) What is the probability that at least one student gets 60 or more heads?
  (b) What is the probability that at least three students get at least 60 heads?

16. Assume that on average there are five raisins per cookie.

(a) What is the probability that in a package of 10 cookies all the cookies have at least one raisin?

(b) How many raisins should each cookie have on average so that the probability in (a) is at least 0.99?

17. Assume that 10% of the population are left-handers. What is the probability that in a class of 40 there are at least three left-handers?

18. Roll a die four times. What is the probability of

(a) getting a pair?

(b) getting three of a kind?

(c) getting four of a kind?

(d) getting two pairs?

(e) four distinct faces?

4
Limit Theorems

4.1 The Law of Large Numbers

Assume that we want to know the mean lifetime of a certain type of battery. A natural way to do that is to pick at random a sample of 100 identical batteries, measure the lifetime for each battery and then compute the average lifetime in our sample. The law of large numbers will show that if the sample is large enough, then the sample average should be close to the true mean with high probability. We now formalize these ideas.

Let $X_1, \ldots, X_n$ be n independent identically distributed (i.i.d.) random variables. These may represent, for instance, the lifetimes of a sample of n batteries. Typically, the distributions of the X_i will not be known. However, we will assume that the mean and the variance exist (but are not known). We denote the variance and the mean of X_i by μ and σ, respectively.

$$
\begin{aligned}
E(X_1) &= E(X_2) = \cdots = E(X_n) = \mu \text{ and} \\
Var(X_1) &= Var(X_2) = \cdots = Var(X_n) = \sigma^2.
\end{aligned}
$$

We would like to estimate μ. A natural estimator for μ is

$$
\bar{X} = \frac{X_1 + X_2 + \cdots + X_n}{n}.
$$

That is, we estimate the true mean μ by using the average over the sample $\bar{X}$. Note that $\bar{X}$ is a random variable whose value varies with the sample over which we are averaging. We start by computing the mean and the variance of $\bar{X}$ as functions of μ and σ. Recall that the expectation is a linear operator.

The Expectation is Linear

Let a_i, $1 \leq i \leq n$, be a sequence of real numbers. Let X_i, $1 \leq i \leq n$, be a sequence of random variables defined on the same sample space. Then

$$E(a_1X_1 + a_2X_2 + \cdots + a_nX_n) = a_1E(X_1) + a_2E(X_2) + \cdots + a_nE(X_n).$$

Note that by the linearity of the expectation,

$$E(X_1 + X_2 + \cdots + X_n) = E(X_1) + E(X_2) + \cdots + E(X_n) = \mu + \mu + \cdots + \mu = n\mu.$$

Again by the linearity of the expectation,

$$E(\bar{X}) = E\left(\frac{X_1 + X_2 + \cdots + X_n}{n}\right) = \frac{1}{n}E(X_1 + X_2 + \cdots + X_n) = \frac{1}{n}n\mu = \mu.$$

That is, the expected value of $\bar{X}$ is the same as the expected value of each random variable X_i. We now compute the variance of $\bar{X}$ in order to investigate the dispersion of $\bar{X}$. We first recall an important property of the variance.

Variance of a Sum of INDEPENDENT Random Variables

Let a_i, $1 \leq i \leq n$, be a sequence of real numbers. Let X_i, $1 \leq i \leq n$, be a sequence of independent random variables defined on the same sample space. Then

$$Var(a_1X_1 + a_2X_2 + \cdots + a_nX_n)$$
$$= a_1^2 Var(X_1) + a_2^2 Var(X_2) + \cdots + a_n^2 Var(X_n).$$

Let $X_1, X_2, \ldots, X_n$ be i.i.d. random variables. We start by computing:

$$
\begin{aligned}
Var(X_1 + X_2 + \cdots + X_n) &= Var(X_1) + Var(X_2) + \cdots + Var(X_n) \\
&= \sigma^2 + \sigma^2 + \cdots + \sigma^2 = n\sigma^2.
\end{aligned}
$$

$$
\begin{aligned}
Var(\bar{X}) &= Var\left(\frac{X_1 + X_2 + \cdots + X_n}{n}\right) = \frac{1}{n^2}Var(X_1 + X_2 + \cdots + X_n) \\
&= \frac{1}{n^2}n\sigma^2 = \frac{\sigma^2}{n}.
\end{aligned}
$$

We summarize these results below.

Expected Value and Variance of the Sample Average

Let $X_1, X_2, \ldots, X_n$ be independent and identically distributed random variables with mean μ and variance σ^2. Then,

$$E(\bar{X}) = E(\frac{X_1 + X_2 + \cdots + X_n}{n}) = \mu$$

and

$$Var(\bar{X}) = Var(\frac{X_1 + X_2 + \cdots + X_n}{n}) = \frac{\sigma^2}{n}.$$

That is, the expected value of $\bar{X}$ is μ and its distribution is more and more concentrated around μ as the sample size n increases.

The variance of $\bar{X}$ is one measure of the distance between $\bar{X}$ and its mean μ. Observe that the variance of $\bar{X}$ converges to 0 as n goes to infinity. In this sense this means that $\bar{X}$ converges to μ as n goes to infinity. In other words, we have justified mathematically the natural idea of taking the sample average to estimate the mean of the distribution.

Example 1. Assume that we use a sample of 100 identical batteries to estimate the lifetime of a battery. Denote the mean and standard deviation of the lifetime distribution by μ and σ respectively. What are the mean and standard deviation of the sample average $\bar{X}$?

According to the formula above

$$E(\bar{X}) = \mu \text{ and } Var(\bar{X}) = \sigma^2/100.$$

Thus, $SD(\bar{X}) = \sigma/10$. That is, the distribution of $\bar{X}$ is 10 times more concentrated than the distribution of X_1.

Since the variance of $\bar{X}$ goes to 0 as the sample size n goes to infinity, we know that $\bar{X}$ approaches μ. But this is not very precise. For instance, it would be more useful to be able to say that $\bar{X}$ is within 0.1 of μ with probability 0.95. We are now going to work towards this goal.

Markov's Inequality

Let $X \geq 0$ be a positive random variable with mean μ. Then, for any $b > 0$,

$$P(X \geq b) \leq \frac{\mu}{b}.$$

Example 2. Find a bound on the probability that a positive random variable will be larger than 10 times its mean.

We want a bound on $P(X > 10\mu)$. We use Markov's inequality with $b = 10\mu$ to get

$$P(X > 10\mu) \leq \frac{\mu}{10\mu} = \frac{1}{10}.$$

What is interesting here is that for *any* positive random variable X (that has a mean) this probability is bound by 0.1.

We now prove Markov's inequality for a continuous random variable. The proof for a discrete random variable is very similar. Let f be the density of X. Then

$$E(X) = \int_0^\infty xf(x)dx = \int_0^b xf(x)dx + \int_b^\infty xf(x)dx \geq \int_b^\infty xf(x)dx.$$

The preceding inequality holds since X is assumed to be positive. Note that

$$\int_b^\infty xf(x)dx \geq b \int_b^\infty f(x)dx = bP(X \geq b).$$

Thus,

$$P(X \geq b) \leq \frac{E(X)}{b}$$

and this proves Markov's inequality.

A consequence of Markov's inequality is Chebyshev's inequality. The latter gives a bound on the dispersion of a random variable.

Chebyshev's Inequality

Let X be a random variable with mean μ and variance σ^2. Then, for any $b > 0$,

$$P(|X - \mu| \geq b) \leq \frac{\sigma^2}{b^2}.$$

Example 3. Let X be a random variable with mean μ and variance σ^2. Give an upper bound on the probability that X is more than 2σ away from its mean μ.

We want $P(|X - \mu| \geq 2\sigma)$. We use Chebyshev's inequality with $b = 2\sigma$ to get

$$P(|X - \mu| \geq 2\sigma) \leq \frac{\sigma^2}{(2\sigma)^2} = \frac{1}{4}.$$

So an upper bound is 1/4. Again, what is remarkable here is that this bound holds for *any* random variable with a variance. The bound given by Chebyshev may be quite crude. For instance, note that if X is a normal random variable, then the probability of being at least 2σ away from μ is about 0.05, while the bound given by Chebyshev's inequality for the same probability is 0.25.

We now prove Chebyshev's inequality. Let X be a random variable with mean μ and standard deviation σ. Define the random variable

$$Y = (X - \mu)^2.$$

Note that $E(Y) = Var(X) = \sigma^2$. We apply Markov's inequality to Y (which is a positive random variable)

$$P(Y > b^2) \leq \frac{E(Y)}{b^2}.$$

The event $\{Y > b^2\}$ may also be written as the event $\{|X - \mu| > b\}$. Thus,

$$P(Y > b^2) = P(|X - \mu| > b) \leq \frac{E(Y)}{b^2} = \frac{\sigma^2}{b^2}$$

and the proof of Chebyshev's inequality is complete.

We are now ready to state the Law of Large Numbers.

Law of Large Numbers

Let $X_1, X_2, \ldots, X_n$ be a sequence of independent identically distributed random variables with mean μ and variance σ^2. Then, for any $b > 0$,

$$\lim_{n \to \infty} P\left(\left|\frac{X_1 + X_2 + \cdots + X_n}{n} - \mu\right| \geq b\right) = 0.$$

Moreover,

$$P\left(\left|\frac{X_1 + X_2 + \cdots + X_n}{n} - \mu\right| \geq b\right) \leq \frac{\sigma^2}{b^2 n}.$$

What the Law of Large Numbers tells us is that the probability that $\bar{X}$ deviates from μ by an arbitrarily small $b > 0$ goes to 0 as the sample size n goes to infinity.

Example 4. Consider a die whose probability to show a 6 is p. Roll the die n times, for each roll let $X_i = 1$ if the die shows a 6 and $X_i = 0$ otherwise, for $i = 1, \ldots, n$. The random variables $X_1, X_2, \ldots, X_n$ are i.i.d. and have a Bernoulli distribution with probability of success p. Recall that for each i,

$$E(X_i) = 0 \times (1 - p) + 1 \times p = p.$$

Thus, according to the Law of Large Numbers we have that for any $b > 0$

$$\lim_{n \to \infty} P\left(\left|\frac{X_1 + X_2 + \cdots + X_n}{n} - p\right| \geq b\right) = 0.$$

This gives a physical interpretation to the notion of probability. When we say that the probability of a 6 is 1/6, it means that if we roll the die n times the ratio of the number of 6s that appear over n will approach 1/6 as n goes to infinity.

Example 5. Assume we roll a die 3600 times and we get 557 6s. Let p be the probability of getting a 6. Find an interval that contains p with probability 0.95.

We use the Bernoulli random variables defined in Example 4. According to the estimate above, we have

$$P\left(\left|\frac{X_1 + X_2 + \cdots + X_n}{n} - p\right| \geq b\right) \leq \frac{\sigma^2}{b^2 n}.$$

In this case we have $\sigma^2 = p(1 - p)$, $\bar{X} = \frac{557}{3600}$ and $n = 3,600$. Since we do not know p (this is what we are estimating) we do not know σ. However, $p(1 - p)$ is always less than 1/4 for p in [0,1] (graph $p(1 - p)$ as a function of p and you will see why). Therefore,

$$P\left(\left|\frac{X_1 + X_2 + \cdots + X_n}{n} - p\right| \geq b\right) \leq \frac{1}{4b^2 n}.$$

We want

$$\frac{1}{4b^2 n} = 0.05.$$

Using that $n = 3,600$ we get $b = 0.04$. The confidence interval with confidence at least 0.95 is

$$\left(\frac{X_1 + X_2 + \cdots + X_n}{n} - b, \frac{X_1 + X_2 + \cdots + X_n}{n} + b\right)$$
$$= (0.15 - 0.04, 0.15 + 0.04)$$
$$= (0.11, 0.19).$$

Such an interval (with a probability attached to it) is called a *confidence* interval.

The confidence interval above is obtained by using Chebyshev's inequality. This inequality holds for all random variables (that have a variance) and in particular for the most dispersed ones. As a consequence the confidence interval we got above is larger than it could be (its confidence is also larger than 0.95). The Central Limit Theorem that we will see in the next section will give us a narrower interval for 0.95 confidence.

Example 6. This example will introduce a numerical integration method called Monte-Carlo integration. Let g be a continuous function on [0,1]. Also Let U_1,

$U_2, \ldots, U_n$ be a sequence of i.i.d. uniform random variables on [0,1]. Then, according to the Law of Large Numbers

$$\lim_{n\to\infty} \frac{g(U_1) + \cdots + g(U_n)}{n} = E(g(U_1)).$$

Recall that if U_1 has density f, then

$$E(g(U_1)) = \int g(x)f(x)dx.$$

In this case $f(x) = 1$ for x in [0,1] and $f(x) = 0$ otherwise. Thus,

$$E(g(U_1)) = \int_0^1 g(x)dx$$

and

$$\lim_{n\to\infty} \frac{g(U_1) + \cdots + g(U_n)}{n} = \int_0^1 g(x)dx.$$

In other words, the average $\frac{g(U_1)+\cdots+g(U_n)}{n}$ approaches $\int_0^1 g(x)dx$ as n goes to infinity. For instance, we take $g(x) = x$, $n = 10$ and use the random numbers: 0.382, 0.101, 0.596, 0.885, 0.899, 0.958, 0.014, 0.407, 0.863, 0.139. We get

$$\frac{g(U_1) + \cdots + g(U_n)}{n} = \frac{U_1 + \cdots + U_n}{n} = 0.52.$$

Thus, 0.52 is the approximation we get for the integral

$$\int_0^1 g(x)dx = \int_0^1 x dx = 0.5.$$

We now prove the Law of Large Numbers. We apply Chebyshev's inequality to the random variable $\bar{X}$. By using that $E(\bar{X}) = \mu$ and $Var(\bar{X}) = \sigma^2/n$, we get for any $b > 0$

$$P(|\bar{X} - \mu| \geq b) \leq \frac{\sigma^2}{nb^2}.$$

As n goes to infinity the right-hand side converges to 0 and this proves the Law of Large Numbers.

Exercises

1. Assume that $X_1, \ldots, X_n$ is a sequence of i.i.d. random variables with mean 3 and standard deviation 2.
   (a) What is the mean of $X_1 + X_2 + \cdots + X_n$?
   (b) What is the mean of $\bar{X}$?

(c) What is the standard deviation of $X_1 + X_2 + \cdots + X_n$?

(d) What is the standard deviation of $\bar{X}$?

2. Find an upper bound on the probability that a positive random variable is 100 times larger than its mean.

3. Assume that the random variable X has mean 3 and standard deviation 2.

(a) Find an upper bound on the probability that X is at least 3σ away from its mean.

(b) Find an upper bound on the probability that X is larger than 11.

4. Let U be uniform in $[0,1]$.

(a) Compute the probability that U is at least σ away from its mean.

(b) Use Chebyshev's inequality to give an upper bound on the probability that U is at least σ away from its mean.

5. Let S be a binomial random variable with $n = 10$ and $p = 0.2$.

(a) Compute the probability that S is at least 2σ away from its mean.

(b) Use Chebyshev's inequality to give an upper bound on the probability that S is at least 2σ away from its mean.

6. (a) Use Monte-Carlo integration to estimate $\int_0^1 e^{-x^2} dx$.

(b) Use the normal table to check the accuracy of the estimate in (a).

7. It is assumed that each line of a given document has a mean of 15 words.

(a) Find an upper bound on the probability that a given line has 30 words or more.

(b) Assume that the standard deviation is $\sigma = 3$. Find a better upper bound for the event in (a).

8. Consider a random variable X such that $P(X = -1) = P(X = 1) = 1/2$.

(a) Compute $E(X)$.

(b) Compute $Var(X)$.

(c) Compute $P(|X - \mu| \geq 1)$.

(d) Show that Chebyshev's inequality is an equality for $P(|X - \mu| \geq 1)$. (This shows that Chebyshev's inequality may not be improved if it is to hold for all random variables with a variance).

9. Show that if p belongs to $[0,1]$ then $p(1 - p) \leq 1/4$.

4.2 Central Limit Theorem

Consider a sequence $X_1, X_2, \ldots, X_n$ of i.i.d. random variables with mean μ and standard deviation σ. We have seen that $\bar{X}$, the sample average, has mean μ and standard deviation $\sigma/\sqrt{n}$. This shows that $\bar{X}$ has a distribution that is more and more concentrated around the mean μ. Actually a lot more is true: the distribution of $\bar{X}$ approaches a normal distribution with mean μ and standard deviation $\sigma/\sqrt{n}$.

Central Limit Theorem

Let $X_1, X_2, \ldots, X_n$ be a sequence of independent identically distributed random variables with mean μ and variance σ^2. Then the distribution of $\bar{X} = \frac{X_1 + X_2 + \cdots + X_n}{n}$ approaches a normal distribution in the following sense. For any $a < b$,

$$\lim_{n \to \infty} P\left(a < \frac{\bar{X} - \mu}{\sigma/\sqrt{n}} < b\right) = P(a < Z < b)$$

where Z has a standard normal distribution. Equivalently, the sum $S = X_1 + X_2 + \cdots + X_n$ also approaches a normal distribution. That is,

$$\lim_{n \to \infty} P\left(a < \frac{S - n\mu}{\sigma\sqrt{n}} < b\right) = P(a < Z < b)$$

We will sketch a proof of the Central Limit Theorem in 7.2. The remarkable fact of this result is that it does not matter what the distribution of the X_i is (provided it has a variance); when we average or sum many i.i.d. random variables we get a normal distribution. The Central Limit Theorem (CLT) shows why the normal distribution is so crucial in probability. We illustrate this theorem with the histograms of $\bar{X}$ for $n = 1$, $n = 3$ and $n = 5$.

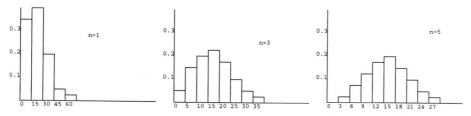

One can see above that when we average even a few random variables there is a departure from the original shape and there is a tendency towards the bell-like shape.

Note that in order to apply the CLT to a random variable Y (we have two choices for Y: $\bar{X}$ and S) we need to standardize it. The CLT states that the distribution of

$$\frac{Y - E(Y)}{SD(Y)}$$ approaches the distribution of Z.

In particular, if $Y = S = X_1 + X_2 + \cdots + X_n$, then $E(S) = n\mu$ and $Var(S) = n\sigma^2$. Thus, the distribution of

$$\frac{S - n\mu}{\sigma\sqrt{n}}$$ approaches the distribution of Z.

On the other hand if $Y = \bar{X}$, then we use that $E(\bar{X}) = \mu$ and that $SD(\bar{X}) = \sigma/\sqrt{n}$ to show that the distribution of

$$\frac{\bar{X} - \mu}{\sigma/\sqrt{n}} \text{ approaches the distribution of } Z.$$

Example 1. We will show that the normal approximation to the binomial is a particular case of the CLT. Consider a binomial random variable S with parameters n and p. Then, S can be written as a sum of n Bernoulli random variables X_i with probability of success p. We have $E(X_i) = p$ and $Var(X_i) = p(1 - p)$. Since the X_i are i.i.d. we may apply the CLT to the sum of X_i to get that the distribution of

$$\frac{S - np}{\sqrt{np(1 - p)}} \text{ approaches the distribution of } Z.$$

This is what the normal approximation to the binomial says. Recall that because we are using a continuous random variable to approach a discrete one, we also enlarge the interval by 1/2 on both sides. That is, we replace $P(a \leq S \leq b)$ by $P(a - 1/2 \leq S \leq b + 1/2)$ before applying the Central Limit Theorem.

Example 2. Toss a fair coin. Each time it lands on heads you win \$1, each time it lands on tails I win \$1. What is the probability that after 100 tosses you will be winning at least \$10?

Let T be your winnings after 100 tosses. We may write T as the sum of the winnings for each toss:

$$T = X_1 + \cdots + X_{100}$$

where X_i (for $1 \leq i \leq 100$) is 1 with probability 1/2 or -1 with probability 1/2. For each i we have that

$$E(X_i) = 1/2 \times 1 + 1/2 \times (-1) = 0.$$

We also have that

$$E(X_i^2) = 1/2 \times (1)^2 + 1/2 \times (-1)^2 = 1.$$

Since $E(X_i) = 0$, $Var(X_i) = E(X_i^2) = 1$. The random variables X_i are i.i.d. so we may use the CLT to get that the distribution of

$$\frac{T - n \times 0}{1\sqrt{100}} \text{ approaches the distribution of } Z.$$

Thus,

$$P(T \geq 10) = P\left(\frac{T}{10} \geq 1\right) \sim P(Z \geq 1) = 0.16.$$

So the probability that you are ahead by at least \$10 is about 0.16. Note that with the same reasoning we get that the probability that you are ahead by at least \$20 after 100 bets is about $P(Z \geq 2) = 0.02$. This is rather unlikely and if this happens one may start to get suspicious about the fairness of the coin.

Example 3. Assume we roll a die 3600 times and we get 557 6s. Let p be the probability of getting a 6. Use the CLT to find an interval that contains p with probability 0.95.

This is the same question as in Example 5 in Section 4.1. We now have the CLT at our disposal so we will use it. Let $X_i = 1$ if the die shows a 6 and $X_i = 0$ otherwise, for $i = 1, \ldots, n$. The random variables $X_1, X_2, \ldots, X_n$ are i.i.d. and have a Bernoulli distribution with probability of success p. Recall that for each i we have

$$E(X_i) = p \text{ and } Var(X_i) = p(1 - p).$$

We want c so that

$$P(|\bar{X} - p| < c) = 0.95.$$

According to the CLT the distribution of

$$\frac{\bar{X} - p}{\sqrt{p(1 - p)}/\sqrt{n}} \text{ approaches the distribution of } Z.$$

Thus,

$$P\left(\frac{|\bar{X} - p|}{\sqrt{p(1 - p)}/\sqrt{n}} < \frac{c}{\sqrt{p(1 - p)}/\sqrt{n}}\right) \sim P\left(|Z| < \frac{c}{\sqrt{p(1 - p)}/\sqrt{n}}\right).$$

We use the normal table to get that

$$\frac{c}{\sqrt{p(1 - p)}/\sqrt{n}} = 1.96.$$

Thus,

$$c = 1.96\sqrt{p(1 - p)}/\sqrt{n}.$$

Since we do not know p (this is what we are estimating) we use that $\sqrt{p(1 - p)} \leq 1/2$ to get that

$$c \leq \frac{1.96}{2\sqrt{3600}} = 0.0016.$$

The confidence interval is

$$(557/3600 - 0.0016; \ 557/3600 + 0.0016).$$

Note that this interval is much narrower (and therefore better) than the interval we got in Example 5 in Section 4.1. This is so because CLT is a much stronger result than Chebyshev's inequality. The price to pay is that CLT is much more difficult to prove than Chebyshev's inequality.

Example 4. Assume that we have 25 batteries whose lifetimes are exponentially distributed with mean two hours. If the batteries are used one at a time, with a failed battery replaced immediately by a new one, what is the probability that after 50 hours there is still a working battery?

Let X_i be the lifetime of the ith battery for $i = 1 \ldots 25$. We want to compute

$$P(X_1 + \cdots + X_{25} > 50).$$

The distribution of a sum of exponentially distributed random variables is not exponentially distributed. To solve this question it is easier to use the CLT rather than use the exact distribution of the sum. The CLT applies since we have an i.i.d. sequence of random variables. Recall that for an exponential random variable the mean and the standard deviation are equal. Thus, in this case we have $\mu = \sigma = 2$. According to the CLT the distribution of

$$\frac{X_1 + \cdots + X_{25} - 25\mu}{\sigma\sqrt{25}} \quad \text{approaches the distribution of } Z.$$

We have that

$$P(X_1 + \cdots + X_{25} > 50)$$
$$= P\left(\frac{X_1 + \cdots + X_{25} - 25 \times 2}{2\sqrt{25}} > \frac{50 - 25 \times 2}{2\sqrt{25}}\right) \sim P(Z > 0) = 0.5.$$

So there is a probability of around 50% that the batteries will last at least 50 hours.

Example 5. We continue Example 4. How many batteries should we have so that there is still a working battery after 50 hours with probability 0.9?

This time we are looking for n such that

$$P(X_1 + \cdots + X_n > 50) = 0.9$$

where X_i is the lifetime of the ith battery. Again we use the CLT to get

$$P(X_1 + X_2 + \cdots + X_n > 50) = P\left(\frac{X_1 + X_2 + \cdots + X_n - n\mu}{\sigma\sqrt{n}} > \frac{50 - n\mu}{\sigma\sqrt{n}}\right)$$
$$\sim P\left(Z > \frac{50 - n\mu}{\sigma\sqrt{n}}\right).$$

Since we want the probability above to be 0.9, we use the normal table to get

$$\frac{50 - n\mu}{\sigma\sqrt{n}} = -1.28.$$

We have that $\mu = \sigma = 2$. Set $x = \sqrt{n}$ and we get the quadratic equation

$$x^2 - 1.28x - 25 = 0.$$

The only positive solution is $x = \sqrt{n} = 5.68$, thus the smallest corresponding integer n is 33. That is, we need at least 33 batteries if we want to ensure a working battery after 50 hours, with probability 0.9.

The CLT tells us what the approximate distribution of a sum (or average) of n i.i.d. random variables is when n is large. However, it does *not* say that every distribution is normal! If the sample size n is not large enough, we need to have more information about the specific distribution we are dealing with and we cannot use the CLT. Next we look at such an example.

Example 6. Assume that 6-year old children have a weight with mean 22 kg and SD of 3 kg. What is the probability that in a class of 30 children at least one weighs over 28 kg?

Let $X_1 \ldots X_{30}$ be the weights of the 30 children. We want

$$P(\max(X_1, X_2, \ldots, X_{30}) > 28).$$

It is easier to deal with the complement of this event

$$P(\max(X_1, X_2, \ldots, X_{30}) > 28) = 1 - P(\max(X_1, X_2, \ldots, X_{30}) \leq 28).$$

Since these random variables are independent,

$$
\begin{aligned}
P(\max(X_1, X_2, \ldots, X_{30}) > 28) &= 1 - P(X_1 \leq 28) P(X_2 \leq 28) \ldots P(X_{30} \leq 28) \\
&= 1 - P(X_1 \leq 28)^{30}
\end{aligned}
$$

where the last equality comes from the fact that all these random variables have the same distribution. In order to compute this probability we need to know the distribution of X_1. To decide what distribution is reasonable we should look at the data. If we *assume* for instance that this distribution is normal, then we get

$$P(X_1 \leq 28) = P\left(\frac{X_1 - 22}{3} \leq \frac{28 - 22}{3}\right) = P(Z \leq 5/3).$$

We use the normal table to get $P(X_1 \leq 28) = 0.95$ and therefore

$$P(\max(X_1, X_2, \ldots, X_{30}) > 28) = 1 - P(X_1 \leq 28)^{30} = 0.78.$$

Of course, if the distribution of X_1 is not close to a normal distribution, then this computation may be very far off.

Exercises

1. Assume that you bet \$1 on red 100 times at the roulette (probability of winning \$1 is 18/38).

(a) What are your expected winnings (or losses) after 100 bets?

(b) What is the probability that you are at least $10 ahead after 100 bets?

(c) Compare (b) to Example 2.

2. Assume that we toss a coin 400 times and we get 260 heads.

(a) Give a confidence interval for the probability p of getting heads with confidence 0.99.

(b) Is this a fair coin?

3. A small airplane can take off with a maximum of 2,000 kg (no luggage!). Assume that passengers have a mean weight of 70 kg with a SD of 15 kg.

(a) What is the probability that 25 passengers will overload the plane?

(b) Find the maximum number of passengers that will not overload the plane with probability 0.99.

4. Assume that first graders have a mean height of 100 cm with SD of 8 cm.

(a) What is the probability that the average height in a class of 30 is over 105 cm?

(b) What is the probability that at least one child is more than 105 cm high?

(c) What assumption did you make to answer (b)?

5. How many times should you toss a fair coin in order to get at least 100 heads with probability 0.9?

6. A bank teller takes a mean of two minutes with a standard deviation of 30 seconds to serve a client. Assuming that there is at least one client waiting at all times, what is the probability that the teller will serve at least 25 clients in one hour?

7. The average grade a professor hands out is 80 with SD of 10.

(a) What is the probability that in a class of 50 the average grade is below 75?

(b) How large should the class be so that the average grade is in the interval [75,85] with probability 0.95?

8. Roll a fair die. What is the probability that the sum of the first 100 rolls will be over 300?

9. Let $X_1 \ldots X_n$ be a sequence of i.i.d. random variables with mean 0 and SD 5. Let S be the sum of the X_i.

(a) What is the probability that S exceeds 10 for $n = 100$?

(b) How large should n be so that at least one of the X_i is larger than 10?

(c) What assumption did you make to answer (b).

10. Example 3 shows that a confidence interval with confidence a for a proportion p has length $2c$ where

$$c = \frac{z_a}{2\sqrt{n}}$$

and z_a is such that

$$P(|Z| < z_a) = a.$$

(a) Does c increase or decrease as the sample size n increases?
(b) What is z_a for $a = 0.9$?
(c) Does c increase or decrease as the confidence a increases?

5

Estimation and Hypothesis Testing

5.1 Large Sample Estimation

Confidence interval for a proportion

Example 1. In a political poll of 100 randomly selected voters, 35 expressed support for initiative A. How does one estimate the proportion of voters in the whole population that supports initiative A based on the sample of 100? How much confidence do we have in our estimate?

Let p be the population proportion of voters in favor of A. A natural estimator for p is $\hat{p}$: the sample proportion of voters in favor of A. Note that p is a constant while $\hat{p}$ is a random variable that depends on the particular sample that we have. Let $X_i = 0$ if the ith voter is against A and let $X_i = 1$ otherwise, for $i = 1, 2, \ldots, 100$. Then,

$$\hat{p} = \frac{X_1 + X_2 + \cdots + X_n}{n}.$$

Note that $X_1, X_2, \ldots, X_n$ are Bernoulli random variables and that $P(X_i = 1) = p$, the parameter we want to estimate. We assume that the sample is random and therefore that the X_i are independent. Recall that $E(X_i) = p$ and so by the linearity of the expectation we get

$$E(\hat{p}) = E\left(\frac{X_1 + X_2 + \cdots + X_n}{n}\right) = \frac{1}{n}E(X_1 + X_2 + \cdots + X_n)$$

$$= \frac{1}{n}nE(X_1) = E(X_1) = p.$$

That is, the expected value of $\hat{p}$ is p; $\hat{p}$ is said to be an *unbiased* estimator of p. We also know, by the Law of Large Numbers, that if the observations X_i are i.i.d., then $\hat{p}$ converges to p as the sample size n increases. So at this point we may say that $\hat{p} = \frac{35}{100}$ is an estimate of p. But what confidence do we have in this estimate? In order to answer this question, we will now compute a confidence interval for p based on $\hat{p}$. That is, we would like to find c such that p is in the interval $(\hat{p} - c, \hat{p} + c)$ with probability 0.95. We need the variance of $\hat{p}$. We use that the X_i are independent, that the variance is a quadratic operator and that the variance of a Bernoulli random variable is $p(1 - p)$ to get

$$Var(\hat{p}) = Var\left(\frac{X_1 + X_2 + \cdots + X_n}{n}\right) = \frac{1}{n^2}Var(X_1 + X_2 + \cdots + X_n)$$
$$= \frac{1}{n^2}nVar(X_1) = \frac{p(1 - p)}{n}.$$

We start by writing that c should be such that

$$P(|\hat{p} - p| < c) = 0.95.$$

If the sample size n is large enough we know, by the Central Limit Theorem, that

$$\frac{\hat{p} - p}{\sqrt{p(1 - p)/n}} \text{ has an approximately standard normal distribution.}$$

Hence,

$$P(|\hat{p} - p| < c) = P\left(|\frac{\hat{p} - p}{\sqrt{p(1 - p)/n}}| < \frac{c}{\sqrt{p(1 - p)/n}}\right) = 0.95.$$

Let Z be a standard normal random variable; we get

$$P\left(|Z| < \frac{c}{\sqrt{p(1 - p)/n}}\right) = 0.95.$$

Using the normal table we have that

$$\frac{c}{\sqrt{p(1 - p)/n}} = 1.96.$$

The above equation has two unknowns: n and p. If n is large enough it is reasonable to estimate p by $\hat{p}$. We get that c is approximately

$$c = 1.96\frac{\sqrt{\hat{p}(1 - \hat{p})}}{\sqrt{n}}.$$

Numerically, we get that $c = 0.09$. Therefore, we may say that with confidence 0.95 the population proportion is in the interval

$$(\hat{p} - c, \hat{p} + c) = (0.35 - 0.09, 0.35 + 0.09) = (0.26, 0.44).$$

One interpretation for the confidence interval above is the following. If we take many samples of 100 voters, then 95% of the confidence intervals we get contain p. Of course, we may be unlucky and draw a sample that will yield an interval that does not contain p. This will happen 5% of the time. Note also that the computations above work only for *random* samples. Asking the opinion of your 100 best friends does not work! One way to draw a random sample from a population is to label all the population and then pick labels at random to get a sample. This is more or less what is done with political polls: phone numbers are selected at random to make up a sample. However, not everybody has a phone and so the poorest part of the population is underrepresented with this method. There are many other things to be cautious about regarding samples; see for instance *Introduction to the Practice of Statistics* by Moore and McCabe, Freeman, third edition, 1999.

We now give the general form of a confidence interval for a proportion.

Confidence Interval for a Proportion

Draw a random sample of size n from a large population with unknown proportion p of successes. Let $\hat{p}$ be the sample proportion of successes. Then, for large n,

$$(\hat{p} - c, \hat{p} + c)$$

is a confidence interval with confidence a where

$$c = z_a \frac{\sqrt{\hat{p}(1 - \hat{p})}}{\sqrt{n}}$$

and z_a is such that

$$P(|Z| < z_a) = a$$

and Z is a standard normal distribution.

Example 2. Find a confidence interval with confidence 0.99 for the proportion in Example 1.

The only difference with Example 1 is the level of confidence. This time $a = 0.99$. According to the normal table

$$P(|Z| < 2.57) = .99$$

so

$$c = 2.57 \frac{\sqrt{\hat{p}(1 - \hat{p})}}{\sqrt{n}}.$$

Numerically, we get $c = 0.12$. Therefore, we may say that with confidence 0.99 the population proportion is in the interval

$$(\hat{p} - c, \hat{p} + c) = (0.35 - 0.12, 0.35 + 0.12) = (0.23, 0.47).$$

Note that at the level 0.99 we get a larger confidence interval. We increase the confidence but we lose in precision.

Example 3. How large should a random sample be to get an estimate of the population proportion within 0.01 with confidence 0.95?

We want to know how large n should be in order to get $c = 0.01$. Since the confidence is $a = 0.95$ we get that $z_a = 1.96$. We need to solve n for the equation

$$c = z_a \frac{\sqrt{p(1-p)}}{\sqrt{n}}.$$

Here we do not know $\hat{p}$ so we use the original p in our formula. A little algebra yields

$$n = (\frac{z_a}{c})^2 p(1-p).$$

However, we do not know p. Note that p is in [0,1] and that the function $g(p) = p(1-p)$ has a maximum for $p = 1/2$. Thus,

$$p(1-p) \le 1/4 \text{ for all } p \text{ in } [0, 1].$$

We get that

$$n \le (\frac{z_a}{c})^2 \frac{1}{4}.$$

Numerically, we get

$$n \le 9,604.$$

That is, in order to get a precision of 0.01 with confidence 0.95 we need a sample of the order of 10,000.

Confidence interval for a mean

Example 4. Assume that 500 lamp bulbs have been tested and the average lifetime for this sample has been 562 days. Give a confidence interval at the level 90% for the mean lifetime of this brand of lamp bulb.

We assume that we have a random sample, i.e., if we denote by $X_1, X_2, \ldots, X_{500}$ the 500 lifetimes observed in the sample, then we assume that this is an i.i.d. sequence of random variables. Denote the mean lifetime by μ and the corresponding standard deviation by σ. We want to estimate μ. A natural estimator for μ is the sample average

$$\bar{X} = \frac{X_1 + X_2 + \cdots + X_n}{n}.$$

Recall that

$$E(\bar{X}) = \mu \text{ and } Var(\bar{X}) = \sigma^2/n.$$

That is, $\bar{X}$ is an unbiased estimator of μ. One way to measure the precision of our estimator is to compute $E((\bar{X} - \mu)^2)$. This is one way to measure the distance between the random variable $\bar{X}$ and the constant μ. Since the expected value of $\bar{X}$ is μ, we get by the definition of the variance that

$$E((\bar{X} - \mu)^2) = Var(\bar{X}) = \sigma^2/n.$$

That is, the precision of our estimator $\bar{X}$ increases as the sample size n increases. In order to compute a confidence interval for μ we need c such that

$$P(|\bar{X} - \mu| < c) = 0.9.$$

We standardize the left-hand side

$$P\left(\frac{|\bar{X} - \mu|}{\sigma/\sqrt{n}} < \frac{c}{\sigma/\sqrt{n}}\right) = 0.9.$$

Since we are assuming that the X_i are i.i.d. and n is large, we may use the Central Limit Theorem to get that

$$\frac{\bar{X} - \mu}{\sigma/\sqrt{n}}$$ has an approximately standard normal distribution.

Let Z be a standard normal random variable; then

$$P\left(|Z| < \frac{c}{\sigma/\sqrt{n}}\right) = 0.9.$$

From the normal table we get

$$\frac{c}{\sigma/\sqrt{n}} = 1.64.$$

However, we usually do not know σ. An unbiased estimator for σ^2 is

$$S^2 = \frac{1}{n-1} \sum_{i=1}^{n} (X_i - \bar{X})^2.$$

We will go back to S^2 below. We get that

$$c = 1.64S/\sqrt{n}.$$

Going back to the observations $X_1, X_2, \ldots, X_n$ one may compute S. Assume that in this example $S = 112$ days. Then, $c = 8$. Thus, a confidence interval for the mean lifetime μ of a lamp bulb, at the 90% level, is

$$(562 - 8, 562 + 8) = (556, 570).$$

Note that this is rather precise thanks to the large size of the sample.

Confidence Interval for a Mean

Draw a random sample of size n, $X_1, X_2, \ldots, X_n$, from a large population with unknown mean μ. Let $\bar{X}$ be the sample mean. Let

$$S^2 = \frac{1}{n-1} \sum_{i=1}^{n} (X_i - \bar{X})^2$$

be the sample variance. Then, for large n,

$$(\bar{X} - c, \bar{X} + c)$$

is a confidence interval for μ with confidence a where

$$c = z_a \frac{S}{\sqrt{n}}$$

and z_a is such that
$$P(|Z| < z_a) = a$$

and Z is a standard normal distribution.

Example 5. Find a confidence interval for the mean in Example 4 with confidence 0.95.

We have that
$$c = z_a \frac{S}{\sqrt{n}} = 1.96 \frac{112}{\sqrt{500}} \sim 10.$$

So at the level 0.95 we have the confidence interval

$$(562 - 10, 562 + 10) = (552, 572)$$

for the mean lifetime of a lamp bulb.

We now go back to the sample standard deviation S. We first establish a computational formula for S. We expand the square below to get

$$\sum_{i=1}^{n} (X_i - \bar{X})^2 = \sum_{i=1}^{n} X_i^2 - 2 \sum_{i=1}^{n} X_i \bar{X} + \sum_{i=1}^{n} (\bar{X})^2.$$

Using that $\sum_{i=1}^{n} X_i = n\bar{X}$ we get that

$$\sum_{i=1}^{n} (X_i - \bar{X})^2 = \sum_{i=1}^{n} X_i^2 - 2n(\bar{X})^2 + n(\bar{X})^2 = \sum_{i=1}^{n} X_i^2 - n(\bar{X})^2.$$

Thus, we get the following computational formula for S^2:

$$S^2 = \frac{1}{n-1} \sum_{i=1}^{n} X_i^2 - \frac{n}{n-1} (\bar{X})^2.$$

We use the preceding formula to compute the expected value of S^2. First, recall that

$$E(X^2) = Var(X) + E(X)^2.$$

Hence,

$$E(S^2) = \frac{1}{n-1} \sum_{i=1}^{n} (\sigma^2 + \mu^2) - \frac{n}{n-1}(Var(\bar{X}) + E(\bar{X})^2),$$

$$E(S^2) = \frac{1}{n-1} n(\sigma^2 + \mu^2) - \frac{n}{n-1}(\sigma^2/n + \mu^2) = \sigma^2.$$

This shows that S^2 is an unbiased estimator of σ^2.

Sample Variance

Let $X_1, X_2, \ldots, X_n$ be i.i.d. random variables with mean μ and variance σ^2 from a large population with unknown mean μ. Then, the sample variance

$$S^2 = \frac{1}{n-1} \sum_{i=1}^{n} (X_i - \bar{X})^2$$

is an unbiased estimator of σ^2. Moreover, we have the following computational formula for S^2:

$$S^2 = \frac{1}{n-1} \sum_{i=1}^{n} X_i^2 - \frac{n}{n-1}(\bar{X})^2.$$

Confidence interval for a difference of proportions

Example 6. In a political poll of 100 randomly selected voters, 35 expressed support for initiative A in Boulder. In Colorado Springs in a poll of 200 randomly selected voters, 50 expressed support for initiative A. Find a confidence interval, with confidence 0.9, for the difference between the proportions of supporters of initiative A in Boulder and in Colorado Springs.

Let p_1 and p_2 be the proportions of the population in Boulder and in Colorado Springs that support A, respectively. Let n_1 and n_2 be the sample sizes taken in Boulder and Colorado Springs, respectively. We would like a confidence interval for $p_1 - p_2$ with confidence 0.9. A natural estimator for $p_1 - p_2$ is $\hat{p}_1 - \hat{p}_2$. We would like to find c so that

$$P(|(\hat{p}_1 - \hat{p}_2) - (p_1 - p_2)| < c) = 0.9.$$

In order to find c we need some information regarding the distribution of $\hat{p}_1 - \hat{p}_2$. We know that if the sample size n_1 is large enough, then by the Central Limit Theorem $\hat{p}_1$ is approximately normally distributed. The same holds for $\hat{p}_2$. Since $\hat{p}_1$ and $\hat{p}_2$ are independent one can show that $\hat{p}_1 - \hat{p}_2$ is approximately normally distributed as well. Since $\hat{p}_1$ and $\hat{p}_2$ are unbiased estimators of p_1 and p_2, respectively, we have that

$$E(\hat{p}_1 - \hat{p}_2) = p_1 - p_2.$$

Since $\hat{p}_1$ is the average of n_1 i.i.d. Bernoulli random variables, we have that

$$Var(\hat{p}_1) = \frac{1}{n_1^2} n_1 p_1(1 - p_1) = \frac{p_1(1 - p_1)}{n_1}.$$

Using that $\hat{p}_1$ and $\hat{p}_2$ are independent we get that

$$Var(\hat{p}_1 - \hat{p}_2) = Var(\hat{p}_1) + Var(\hat{p}_2) = \frac{p_1(1 - p_1)}{n_1} + \frac{p_2(1 - p_2)}{n_2}.$$

We are now ready to normalize to get

$$P\left(\left|\frac{(\hat{p}_1 - \hat{p}_2) - (p_1 - p_2)}{\sqrt{\frac{p_1(1-p_1)}{n_1} + \frac{p_2(1-p_2)}{n_2}}}\right| < \frac{c}{\sqrt{\frac{p_1(1-p_1)}{n_1} + \frac{p_2(1-p_2)}{n_2}}}\right) = 0.9.$$

We use that

$$\frac{(\hat{p}_1 - \hat{p}_2) - (p_1 - p_2)}{\sqrt{\frac{p_1(1-p_1)}{n_1} + \frac{p_2(1-p_2)}{n_2}}} \text{ is approximately a standard normal distribution}$$

to get that

$$P\left(|Z| < \frac{c}{\sqrt{\frac{p_1(1-p_1)}{n_1} + \frac{p_2(1-p_2)}{n_2}}}\right) = 0.9$$

where Z is a standard normal random variable. Using the normal table we get

$$\frac{c}{\sqrt{\frac{p_1(1-p_1)}{n_1} + \frac{p_2(1-p_2)}{n_2}}} = 1.64.$$

For n_1 and n_2 large enough we may use $\hat{p}_1$ and $\hat{p}_2$ to approximate p_1 and p_2, respectively. Thus,

$$c \sim 1.64\sqrt{\frac{\hat{p}_1(1 - \hat{p}_1)}{n_1} + \frac{\hat{p}_2(1 - \hat{p}_2)}{n_2}}.$$

In this example we have $n_1 = 100$, $n_2 = 200$, $\hat{p}_1 = 0.35$ and $\hat{p}_2 = 0.25$. Thus, $c = 0.09$. At the level 90% the confidence interval for $p_1 - p_2$ is

$$(\hat{p}_1 - \hat{p}_2 - c, \hat{p}_1 - \hat{p}_2 + c) = (0.01, 0.19).$$

We now summarize the preceding technique.

Confidence Interval for the Difference Between Two Proportions

Draw a random sample of size n_1 from a large population with unknown proportion p_1 of successes and an independent random sample of size n_2 from another population having a proportion p_2 of successes. For large n_1 and large n_2,

$$(\hat{p}_1 - \hat{p}_2 - c, \ \hat{p}_1 - \hat{p}_2 + c)$$

is a confidence interval with confidence a where

$$c = z_a \sqrt{\frac{\hat{p}_1(1 - \hat{p}_1)}{n_1} + \frac{\hat{p}_2(1 - \hat{p}_2)}{n_2}}$$

and z_a is such that

$$P(|Z| < z_a) = a$$

and Z is a standard normal distribution.

Confidence interval for a difference of two means

Example 7. Assume that $n_1 = 500$ lamp bulbs from brand 1 have been tested. The average lifetime for this sample is $\bar{X}_1 = 562$ days; the standard deviation for the sample is $S_1 = 112$. Similarly, $n_2 = 300$ lamp bulbs from brand 2 have been tested. The average lifetime for this sample is $\bar{X}_2 = 551$ days and the standard deviation for the sample is $S_2 = 121$. Give a confidence interval at the level 95% for the difference of mean lifetimes of the two brands of lamp bulb. Is there evidence that brand 1 lasts longer than brand 2?

Let μ_1 and μ_2 be the unknown mean lifetimes of brands 1 and 2, respectively. We would like a confidence interval for $\mu_1 - \mu_2$. We use $\bar{X}_1 - \bar{X}_2$ as an estimator of $\mu_1 - \mu_2$. We want c such that

$$P(|(\bar{X}_1 - \bar{X}_2) - (\mu_1 - \mu_2)| < c) = 0.95.$$

In order to find c we need to know the distribution of the expression above. By the Central Limit Theorem $\bar{X}_1$ and $\bar{X}_2$ are approximately normally distributed if the sample sizes are large enough. If the two samples are independent, then one can show that $\bar{X}_1 - \bar{X}_2$ is also approximately normally distributed. Since $\bar{X}_1$ and $\bar{X}_2$ are unbiased estimators of μ_1 and μ_2 we have that

$$E(\bar{X}_1 - \bar{X}_2) = \mu_1 - \mu_2.$$

Since $\bar{X}_1$ and $\bar{X}_2$ are independent we get

$$Var(\bar{X}_1 - \bar{X}_2) = Var(\bar{X}_1) + Var(\bar{X}_2) = \sigma_1^2/n_1 + \sigma_2^2/n_2.$$

We are now ready to normalize to get

$$P\left(\frac{|(\bar{X}_1 - \bar{X}_2) - (\mu_1 - \mu_2)|}{\sqrt{\sigma_1^2/n_1 + \sigma_2^2/n_2}} < \frac{c}{\sqrt{\sigma_1^2/n_1 + \sigma_2^2/n_2}}\right) = 0.95.$$

By using the normal approximation we get

$$P\left(|Z| < \frac{c}{\sqrt{\sigma_1^2/n_1 + \sigma_2^2/n_2}}\right) = 0.95.$$

According to the normal table, we get

$$\frac{c}{\sqrt{\sigma_1^2/n_1 + \sigma_2^2/n_2}} = 1.96.$$

However, the variances σ_1^2 and σ_2^2 are not known. If n_1 and n_2 are large it is reasonable to use the sample variances S_1^2 and S_2^2 in order to estimate them.

$$c \sim 1.96\sqrt{S_1^2/n_1 + S_2^2/n_2}.$$

Numerically, we get $c \sim 17$. Thus, the confidence interval for $\mu_1 - \mu_2$ at the level 0.95 is

$$(\bar{X}_1 - \bar{X}_2 - c, \bar{X}_1 - \bar{X}_2 + c) = (-6, 28).$$

Note that 0 is in the above interval. This shows that μ_1 and μ_2 could be equal. So there is no evidence that brand 1 lasts longer than brand 2.

Confidence Interval for the Difference between Two Means

Draw a random sample of size n_1 from a large population with unknown mean μ_1 and an independent random sample of size n_2 from another population having mean μ_2. We denote the sample average by $\bar{X}_i$ and the sample standard deviation by S_i, for $i = 1, 2$. For large n_1 and large n_2

$$(\bar{X}_1 - \bar{X}_2 - c, \bar{X}_1 - \bar{X}_2 + c)$$

is a confidence interval with confidence a where

$$c = z_a\sqrt{S_1^2/n_1 + S_2^2/n_2}$$

and z_a is such that

$$P(|Z| < z_a) = a$$

and Z is a standard normal distribution.

Exercises

1. Consider the following scores: 87, 92, 58, 64, 72, 43, 75. Compute the average score $\bar{X}$ and the standard deviation S.

2. The English statistician Karl Pearson once tossed a coin 24,000 times and obtained 12,012 heads. Find a confidence at the level 0.99 for the probability of heads.

3. A poll institute claims that its estimate of a proportion is within 0.02 of the true value with confidence 0.95. How large must the sample be?

4. A poll institute has interviewed 1000 people and gives an estimate of a proportion within 0.01. What is the confidence of this estimate?

5. Of 250 Ponderosa pines attacked with a certain type of beetle, 34 died. Find a confidence interval at the level 0.9 for the proportion of trees that die when attacked by this type of beetle.

6. The heights of 25 6-year old boys average 85 cm with a standard deviation of 5 cm. Find a confidence interval at the level 0.95 for the mean height of a 6-year old boy of that population.

7. A researcher has measured the yields of 40 tomato plants and found the sample average yield per plant to be 5 pounds with a sample standard deviation of 1.7 pounds. Find a confidence interval for the mean yield at the level 0.9.

8. The English statistician Karl Pearson once tossed a coin 24,000 times and obtained 12,012 heads. The English mathematician John Kerrich tossed a coin 10,000 times and obtained 5,067. Find a confidence at the level 0.99 for the difference of the probabilities of heads for the two coins.

9. The same final exam is given to several sections of calculus students. Each professor gets to grade 50 papers taken at random from the pile. Professor A has an average of 75 with a standard deviation of 12. Professor B has an average of 79 with a standard deviation of 8.

   (a) Find a confidence interval for the difference in mean scores between Professors A and B.

   (b) Is there evidence that Professor A is harsher than Professor B?

10. In March a poll indicates that 104 out of 250 voters are in favor of initiative A. In October another (independent of the first one) indicates that 140 out of 300 voters are in favor of initiative A.

   (a) Find a confidence interval for the difference of proportions of voters in favor of initiative A in March and October.

   (b) Based on (a) would you say that there is statistical evidence that support has increased for initiative A?

11. A researcher wants to compare the yield of two varieties of tomatoes. The first variety of 40 tomato plants has a sample average yield per plant of 5 pounds with a sample standard deviation of 1.7 pounds. The second variety of 50 tomato plants has a sample average yield per plant of 4.5 pounds with a sample standard deviation of 1.2 pounds.

(a) Find a confidence interval for the difference in mean yield at the level 0.9.

(b) Based on (a), would you say that variety 1 yields more than variety 2?

12. Consider a random sample $X_1, \ldots, X_n$ of size of a uniform random variable on $[0, a]$. Recall that $E(X_1) = a/2$.

(a) Find an unbiased estimator $\hat{a}$ for a.

(b) Compute $E((\hat{a} - a)^2)$ (this indicates how close $\hat{a}$ is to a).

13. Compute the expected value of

$$\frac{1}{n} \sum_{i=1}^{n} (X_i - \bar{X})^2.$$

5.2   Hypothesis Testing

Testing a proportion

Example 1. A manufacturer claims that he produces strictly less than 5% defective items. A sample of 100 items is taken at random and four are found to be defective. Test the claim of the manufacturer.

We denote the true proportion of defective items by p. The claim of the manufacturer is that $p < 0.05$. We want to test whether this claim holds based on the observations. There are two possible errors. We may reject the claim of the manufacturer although it is true or we may accept the claim of the manufacturer although it is not true. The test we will perform is not symmetric and the errors cannot be both small. We set up the test so that the error we minimize is the one that accepts the claim of the manufacturer although it is not true. This is so because we are testing the manufacturer's claim and he should have the burden of proof. The manufacturer's claim is called the *alternative hypothesis* and it is denoted by H_a. The negation of this claim is called the *null hypothesis* and is denoted by H_0. So the test we would like to perform is

$$H_0 : p \geq 0.05,$$
$$H_a : p < 0.05.$$

It is convenient to have an equality for the null hypothesis. It is always possible to replace an inequality by an equality in the null hypothesis without changing the test. We will actually test

$$H_0 : p = 0.05,$$
$$H_a : p < 0.05.$$

We need to make a decision: reject H_0 (the manufacturer's claim is accepted) or do not reject H_0 (the manufacturer's claim is not accepted). We make this decision based on the observations. We will reject H_0 if the sample proportion is low. We define the *rejection region* to be

$$R = \{\hat{p} < c\}.$$

If the sample proportion $\hat{p}$ is in the rejection region we reject H_0 (the manufacturer's claim is accepted). We determine c by fixing a *significance level* usually denoted by α. The significance level is the error we make when we reject H_0 although H_0 holds. Thus,

$$P(\text{ reject } H_0 | H_0 \text{ is true}) = \alpha.$$

Therefore,

$$\alpha = P(\hat{p} < c | H_0) = P(\hat{p} < c | p = 0.05).$$

We normalize to get

$$\alpha = P\left(\frac{\hat{p} - p}{\sqrt{p(1-p)/n}} < \frac{c - p}{\sqrt{p(1-p)/n}} \middle| p = 0.05\right).$$

If the sample size is large enough we may use the Central Limit Theorem and we have

$$\alpha = P\left(Z < \frac{c - 0.05}{\sqrt{0.05(0.95)/n}}\right)$$

where Z is a standard normal random variable. We set $\alpha = 5\%$ which is a typical value for α. We get by the normal table

$$\frac{c - 0.05}{\sqrt{0.05(0.95)/n}} = -1.64.$$

We have

$$c = 0.05 - 1.64\sqrt{0.05(0.95)/n},$$

so that the rejection region is

$$R = \{\hat{p} < 0.01\}.$$

Since $\hat{p} = 0.04$ for this sample, we do not reject H_0. Our test rejects the claim of the manufacturer at the 0.05 significance level. What this test is telling us is that although we observe 4% of defective items in the sample there is a probability (larger than 5%) that this was due to chance and that the actual proportion of defective items is equal to or larger than 5%.

Hypothesis Testing

The claim you want to test should be your alternative hypothesis and is denoted by H_a. The negation of that claim is called the null hypothesis and is denoted by H_0. The test is determined by the level of significance. This is the probability of making the following error: rejecting H_0 although H_0 is true.

In practice it is better not to fix a significance level but instead to compute the so called *P-value* of the test. We go back to Example 1. We know that the rejection region is $R = \{\hat{p} < c\}$. Instead of fixing α and finding c as we did before, we use for c the observed $\hat{p}$. So in this case the rejection region is $R = \{\hat{p} < 0.04\}$. We now compute the significance associated with this rejection region. This is the P-value

$$P = P(\text{reject} H_0 | H_0 \text{is true}) = P(\hat{p} < 0.04 | p = 0.05).$$

We normalize and use the CLT again to get

$$P = P\left(\frac{\hat{p} - 0.05}{\sqrt{0.05(0.95)/n}} < \frac{0.04 - 0.05}{\sqrt{0.05(0.95)/n}}\right) = P(Z < -0.46) = 0.32.$$

The advantage of computing the P-value is that it provides the test for as many significant levels as we want. For instance, if we want to test the claim at the 5% level, then we see that the rejection region will be smaller than the one associated with the P-value. Therefore the observed $\hat{p}$ will not be in the 5% rejection region and we will not reject H_0 at the 5% level. We will only reject H_0 at a level larger than 0.32. There is no reason why we should accept such a large probability of error. So at any reasonable level (usually smaller than 0.1) we will not reject H_0.

P Value

The P-value of a test is the probability that we reject H_0 although H_0 holds based on the rejection region associated with the sample. For a significance larger than the P-value, the hypothesis H_0 should be rejected. For a significance smaller than the P-value, the hypothesis H_0 should not be rejected.

We summarize the P-value method for testing a proportion.

P Value for Testing a Proportion

Assume we have a large random sample with a proportion $\hat{p}$ of successes. We use this sample to test the true proportion of successes p. Let p_0 be a fixed number in $[0,1]$. For the test,

$$H_0 : p = p_0,$$
$$H_a : p < p_0,$$

and a sample size n large enough, the P-value is

$$P = P\left(Z < \frac{\hat{p} - p_0}{\sqrt{p_0(1 - p_0)/n}}\right)$$

where $\hat{p}$ is the sample proportion and n is the size of the random sample. For the test

$$H_0 : p = p_0,$$
$$H_a : p > p_0,$$

and a sample size n large enough, the P-value is

$$P = P\left(Z > \frac{\hat{p} - p_0}{\sqrt{p_0(1 - p_0)/n}}\right).$$

Example 2. Assume that in a poll, candidate A got 121 votes in a random sample of 1,000. Candidate A claims that more than 10% of the voters are in his favor. Test his claim.

Let p be the proportion of voters in favor of candidate A. The alternative hypothesis should be $p > 0.1$ since this is the claim we want to test. Therefore, the test is

$$H_0 : p = 0.1,$$
$$H_a : p > 0.1.$$

The P-value is

$$P = P\left(Z > \frac{\hat{p} - p_0}{\sqrt{p_0(1 - p_0)/n}}\right) = P\left(Z > \frac{0.121 - 0.1}{\sqrt{0.1(0.9)/(1,000)}}\right)$$
$$= P(Z > 2.21) = 0.01.$$

Since $P < 0.05$, at the level 5% we reject H_0: there is statistical evidence supporting the claim of candidate A.

Testing a mean

Example 3. A manufacturer of lamp bulbs claims that the mean lifetime of his lamp bulbs is 1,000 hours. The average lifetime in a sample of 200 bulbs is 1,016 hours with a standard deviation of 102 hours. Test the claim of the manufacturer.

Let μ be the true lifetime mean of a lamp bulb. The claim of the manufacturer is that $\mu > 1,000$. So this should be our alternative hypothesis. Therefore, the test is going to be

$$H_0 : \mu = 1,000,$$
$$H_a : \mu > 1,000.$$

To take our decision on μ we use $\bar{X}$: the average lifetime in the sample. The rejection region is of the type

$$R = \{\bar{X} > c\}.$$

We compute the P-value, that is we take $c = 1,016$.

$$P = P(\bar{X} > 1,016|\mu = 1,000).$$

If the sample size is large enough we may use the CLT to get

$$P = P\left(\frac{\bar{X} - 1,000}{S/\sqrt{n}} > \frac{1,016 - 1,000}{S/\sqrt{n}}\right) \sim P(Z > 2.21) = 0.01.$$

So at any level larger than 0.01 we reject H_0. In particular at the standard level 0.05 we reject H_0. There is statistical evidence supporting the manufacturer's claim.

P Value for Testing a Mean

Assume we have a large random sample with average $\bar{X}$ and standard deviation S. We would like to use this sample to test the true mean of the population μ. Let μ_0 be a fixed number. For the test

$$H_0 : \mu = \mu_0,$$
$$H_a : \mu < \mu_0,$$

the P-value is

$$P = P\left(Z < \frac{\bar{X} - \mu_0}{S/\sqrt{n}}\right).$$

For the test

$$H_0 : \mu = \mu_0,$$
$$H_a : \mu > \mu_0,$$

the P-value is

$$P = P\left(Z > \frac{\bar{X} - \mu_0}{S/\sqrt{n}}\right).$$

Example 4. A farmer is supposed to deliver to a grocery store chickens that weigh 3 pounds on average. A random sample of 100 chicken has an average of 46 ounces and a standard deviation of 5 ounces. The grocery store claims that the chickens are on average under 3 pounds. Test the claim of the store.

The claim we want to test is $\mu < 48$. This should be our alternative hypothesis. So we perform the test

$$H_0 : \mu = 48,$$
$$H_a : \mu < 48.$$

We compute the P-value.

$$P = P\left(Z < \frac{\bar{X} - \mu_0}{S/\sqrt{n}}\right) = P\left(Z < \frac{46 - 48}{5/\sqrt{100}}\right) = P(Z < -4).$$

This P-value is practically 0. So at any reasonable level (1%, 5% or 10%) we should reject the null hypothesis. There is strong statistical evidence to support the claim of the grocery store.

Testing two proportions

Example 5. In a poll of 1,000 voters, candidate A got 42% of the votes in Colorado Springs. In Boulder he got 39% of the votes in a poll of 500 voters. Is the support of candidate A larger in Colorado Springs than in Boulder?

Let p_1 and p_2 be respectively the true proportions of voters in favor of A in Colorado Springs and in Boulder. We want to test whether $p_1 > p_2$. So we want to perform the test

$$H_0 : p_1 = p_2,$$
$$H_a : p_1 > p_2.$$

It is convenient to observe that this test can be expressed as a one-parameter test by writing it as

$$H_0 : p_1 - p_2 = 0,$$
$$H_a : p_1 - p_2 > 0.$$

The rejection region is

$$R = \{\hat{p}_1 - \hat{p}_2 > c\}.$$

We compute the P-value for this test. That is, we take $c = 0.03$.

$$P = P(\hat{p}_1 - \hat{p}_2 > 0.03 | p_1 - p_2 = 0).$$

We use that

$$E(\hat{p}_1 - \hat{p}_2) = p_1 - p_2 \text{ and } Var(\hat{p}_1 - \hat{p}_2) = p_1(1 - p_1)/n_1 + p_2(1 - p_2)/n_2.$$

For n_1 and n_2 large, and if the two random samples are independent, we may use the CLT to get

$$P = P\left(Z > \frac{0.03}{\sqrt{p_1(1 - p_1)/n_1 + p_2(1 - p_2)/n_2}} \middle| p_1 - p_2 = 0 \right).$$

We need to estimate p_1 and p_2 in the expression. Given that we are assuming that $p_1 = p_2$ we use the pooled estimate

$$\hat{p} = \frac{n_1\hat{p}_1 + n_2\hat{p}_2}{n_1 + n_2}.$$

Thus,

$$P = P\left(Z > \frac{0.03}{\sqrt{\hat{p}(1 - \hat{p})(1/n_1 + 1/n_2)}} \right).$$

Numerically, we get $\hat{p} = 0.41$ and that $P = P(Z > 1.11) = 0.13$. At the 5% (or 10%) level we do not reject H_0. There is no evidence that the support of candidate A is larger in Colorado Springs than in Boulder.

P Value for Testing Two Proportions

We have two independent random samples of size n_1 and n_2 respectively from two distinct populations. Let p_1 and p_2 be respectively the true proportions of successes in populations 1 and 2. Let $\hat{p}_1$ and $\hat{p}_2$ be the corresponding sample proportions. For the test

$$H_0 : p_1 = p_2,$$
$$H_a : p_1 < p_2,$$

the P-value is

$$P = P\left(Z < \frac{\hat{p}_1 - \hat{p}_2}{\sqrt{\hat{p}(1 - \hat{p})(1/n_1 + 1/n_2)}} \right)$$

where

$$\hat{p} = \frac{n_1 \hat{p}_1 + n_2 \hat{p}_2}{n_1 + n_2}.$$

For the test

$$H_0 : p_1 = p_2,$$
$$H_a : p_1 > p_2,$$

the P-value is

$$P = P\left(Z > \frac{\hat{p}_1 - \hat{p}_2}{\sqrt{\hat{p}(1 - \hat{p})(1/n_1 + 1/n_2)}} \right).$$

Testing two means

Example 6. We test 200 lamp bulbs of manufacturer A and find that the sample average is 1,052 hours and the standard deviation average is 151. We test 100 lamps of manufacturer B and find that the sample average is 980 hours with a standard deviation of 102. Test the claim that lamp bulbs from A last longer than lamp bulbs from B.

Let μ_1 and μ_2 be respectively the mean lifetimes of the lamp bulbs from manufacturers A and B. We want to test whether $\mu_1 > \mu_2$. This is our alternative hypothesis. We perform the test

$$H_0 : \mu_1 = \mu_2,$$
$$H_a : \mu_1 > \mu_2.$$

We rewrite the test as a one-parameter test

$$H_0 : \mu_1 - \mu_2 = 0,$$
$$H_a : \mu_1 - \mu_2 > 0.$$

Let n_1 and n_2 be the sample sizes from A and B, respectively. We denote the sample averages from A and B by $\bar{X}_1$ and $\bar{X}_2$, respectively and the sample standard deviations by S_1 and S_2. The rejection region is of the type

$$R = \{\bar{X}_1 - \bar{X}_2 > c\}.$$

To compute the P-value we take $c = 1,052 - 980 = 72$.

$$P = P(\bar{X}_1 - \bar{X}_2 > 72 | \mu_1 - \mu_2 = 0).$$

Assuming the sample sizes are large enough and that the two random samples are independent, we get by the Central Limit Theorem that

$$P \sim P(Z > \frac{72}{\sqrt{S_1^2/n_1 + S_2^2/n_2}}) = P(Z > 4.87).$$

This is an extremely small P-value. At any reasonable level we reject H_0. There is strong statistical evidence supporting the claim that lamp bulbs from A last longer than lamp bulbs from B. In order to estimate how much longer lamp bulbs from brand A last, we may compute a confidence interval. For instance with 95% confidence we get the following confidence interval for $\mu_1 - \mu_2$:

$$(\bar{X}_1 - \bar{X}_2 - c, \bar{X}_1 - \bar{X}_2 + c)$$

where

$$c = z_a \sqrt{S_1^2/n_1 + S_2^2/n_2}.$$

We have $z_a = 1.96$ and $c \sim 29$. So the confidence interval for $\mu_1 - \mu_2$ with 0.95 confidence is (43, 101).

P Value for Testing Two Means

Draw a random sample of size n_1 from a large population with unknown mean μ_1 and an independent random sample of size n_2 from another population having mean μ_2. We denote the sample average by $\bar{X}_i$ and the sample standard deviation by S_i, for $i = 1, 2$. For large n_1 and large n_2, to test

$$H_0 : \mu_1 = \mu_2,$$
$$H_a : \mu_1 > \mu_2,$$

the P-value is

$$P = P\left(Z > \frac{\bar{X}_1 - \bar{X}_2}{\sqrt{S_1^2/n_1 + S_2^2/n_2}}\right).$$

To test

$$H_0 : \mu_1 = \mu_2,$$
$$H_a : \mu_1 < \mu_2,$$

the P-value is

$$P = P\left(Z < \frac{\bar{X}_1 - \bar{X}_2}{\sqrt{S_1^2/n_1 + S_2^2/n_2}}\right).$$

A few remarks

This chapter is a very short introduction to statistics. The confidence intervals and hypothesis testing we have performed all assume that the samples are *random* and *large*. However, it is possible to analyze small samples (this might be necessary in areas like medicine for which one does not always control the sample sizes) by using different techniques. We will give two such examples in the next section.

For hypothesis testing we have concentrated on one error: rejecting the null hypothesis when it is true. This is called a type I error. There is another possible error which is not rejecting the null hypothesis when it is not true. This is called a type II error. As we have seen it is the type I error that determines a test. However, two different tests with the same type I error may be compared by computing the type II errors. The test with the lower type II error (given a type I error) is the better one.

Exercises

1. Assume that in a random sample of 1,000 items 40 are defective. The manufacturer claims that less than 5% of the items are defectives.

(a) Test the claim of the manufacturer at the level 10%.

(b) Compare the conclusion of (a) to the conclusion of Example 1.

2. A pharmaceutical company claims that its new drug is more efficient than the existing one that cures about 70% of the cases treated. In a random sample of 96 patients 81 were cured by the new drug.

(a) What test should the company perform to prove its point?

(b) Perform the test stated in (a) and draw a conclusion.

3. Two drugs are compared. Drug A was given to 31 patients and 25 recovered. Drug B was given to 42 patients and 32 recovered. Is there evidence that drug A is more effective than drug B?

4. Pesticide A killed 15 of 35 cockroaches and pesticide B killed 20 of 35 cockroaches. Compare the two pesticides.

5. Radon detectors from a company are tested. Each detector is exposed to 100 standard units of radon. For a sample of 25 detectors, the average reading was 97 with a standard deviation of 8. Is there evidence that the detectors are undermeasuring radon levels?

6. The heights of 25 6-year old boys from vegetarian families average 85 cm with a standard deviation of 5 cm. The average height for the general population of 6-year old boys is 88 cm. Is there evidence that children from vegetarian families are not as tall as children in the general population?

7. With pesticide A in a sample of 40 plants the average yield per plant is 5 pounds with a sample standard deviation of 1.7 pound. Using pesticide B in a sample of 30 plants the average yield is 4.5 pounds with a standard deviation of 1.5 pound. Compare the two pesticides.

8. To study the effect of a drug on pulse rate, the available subjects were divided at random in two groups of 30 persons each. The first group was given the drug. The second group was given a placebo. The treatment group had an average pulse rate of 67 with a standard deviation of 8. The placebo group had an average pulse rate of 71 with a standard deviation of 10. Test the effectiveness of the drug.

9. An aptitude test is given in 6th grade. The 150 boys average 75 with a standard deviation of 10 and the 160 girls average 87 with a standard deviation of 8. Test the claim that girls are in average at least 10 points above boys.

10. A coin is tossed 12 times and 9 tails are observed. Is there evidence that the coin is biased? (The sample is too small to use the CLT but you may use the binomial distribution for the number of heads in 12 tosses).

5.3 Small Samples

In the previous two sections we have used the Central Limit Theorem to get confidence intervals and perform hypothesis testing. Usually the CLT may be safely applied for random samples of size 25 or larger. In this section, we will see two alternatives for smaller sample sizes.

If the population is normal

Assuming we have a random sample of size n from a normal population, then it is possible to compute the exact distribution of the normalized sample mean (recall that the CLT only gives an approximate distribution). We now state this result precisely.

Distribution of the Sample Mean

Assume that $X_1, X_2, \ldots, X_n$ are observations from a random sample taken in a NORMAL population. Let μ be the true mean. Let $\bar{X}$ and S be respectively the sample average and the sample standard deviation. Then

$$\frac{\bar{X} - \mu}{S/\sqrt{n}}$$

follows a student distribution with $n - 1$ degrees of freedom. A Student random variable with r degrees of freedom will be denoted by $t(r)$.

Student distributions are very similar to the normal standard distributions: they are bell shaped and symmetric around the y axis. The only difference is that the tails of the Student distribution are larger than the tails of the standard normal distribution. That is, the probability of an extreme value is higher for a Student distribution than for a standard normal distribution. However, as the number of degrees of freedom increases, Student distributions are closer and closer to the standard normal distribution (as it should be according to the CLT).

The graph below compares a Student distribution with 2 degrees of freedom with a standard normal distribution. One sees for instance that it is a lot more likely for $t(2)$ to be larger than 3 than it is for Z to be larger than 3.

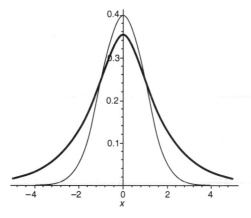

Example 1. Assume that the weights of five 9-year old boys are 25, 28, 24, 26 and 24 kilograms in a certain population. Find a confidence interval for the mean weight of 9-year old boys in that population.

We first need to compute the sample average and standard deviation.

$$\bar{X} = \frac{2 \times 24 + 25 + 26 + 28}{5} = 25.4.$$

We have the following computational formula for S^2,

$$S^2 = \frac{1}{n-1} \sum_{i=1}^{n} X_i^2 - \frac{n}{n-1} (\bar{X})^2.$$

We compute the sum of the squares first,

$$\sum_{i=1}^{n} X_i^2 = 3237$$

and we get

$$S^2 = \frac{1}{4} 3237 - \frac{5}{4} 25.4^2 = 2.8.$$

As always we use the sample average to estimate the true mean μ. Thus, we look for c such that

$$P(|\bar{X} - \mu| < c) = 0.9.$$

We normalize to get

$$P\left(\frac{|\bar{X} - \mu|}{S/\sqrt{n}} < \frac{c}{S/\sqrt{n}} \right) = 0.9.$$

At this point we need the distribution of $\frac{\bar{X} - \mu}{S/\sqrt{n}}$. The sample is much too small to invoke the CLT. However, it may be reasonable to *assume* that the weight is normally distributed. In that case $\frac{\bar{X} - \mu}{S/\sqrt{n}}$ follows a Student distribution with 4 degrees of freedom $t(4)$. Thus,

$$P\left(|t(4)| < \frac{c}{S/\sqrt{n}} \right) = 0.9.$$

We now use the t table to get that

$$\frac{c}{S/\sqrt{n}} = 2.13.$$

Solving for c we get $c = 1.6$. Thus, the confidence interval for the true mean weight of 9-year olds is

$$(\bar{X} - c, \bar{X} + c) = (25.4 - 1.6, 25.4 + 1.6) = (23.8, 27).$$

If we were using the standard normal distribution, then we would have had

$$\frac{c}{S/\sqrt{n}} = 1.64$$

instead of 2.13 and the confidence interval would have been narrower. However, the sample size is too small to invoke the CLT in this case. Next we summarize the method to find a confidence interval for the mean.

Confidence Interval for a Mean

Draw a random sample of size n, $X_1, X_2, \ldots, X_n$, from a NORMAL population with unknown mean μ. Let $\bar{X}$ be the sample mean. Let S^2 be the sample variance. Then,

$$(\bar{X} - c, \bar{X} + c)$$

is a confidence interval for μ with confidence a where

$$c = t_a \frac{S}{\sqrt{n}}$$

and t_a is such that

$$P(|t(n-1)| < t_a) = a$$

and $t(n-1)$ is a Student distribution with $n-1$ degrees of freedom.

Example 2. Assume that the average salary of six associate professors at a certain Mathematics department is 45, 000 with a standard deviation of 2, 500. Assume that the national average salary for associate professors in mathematics is 51,000. Is there evidence that professors are earning less than the national average at this institution?

We would like to test the claim that the salary at this institution is smaller than the national average. So the alternative hypothesis H_a is $\mu < 51$. We test

$$H_0 : \mu = 51,$$
$$H_a : \mu < 51.$$

We compute the P-value for this test.

$$P = P(\bar{X} < 45 | \mu = 51) = P\left(\frac{\bar{X} - 51}{S} < \frac{45 - 51}{S}\right).$$

Assuming that the salaries are normally distributed we get that $\frac{\bar{X}-51}{S}$ follows a Student distribution $t(5)$. Thus,

$$P = P(t(5) < -2.4) \sim 0.03.$$

At the 5% level the null hypothesis is rejected and we see that there is evidence that associate professors are underpaid at that institution.

P Value for Testing a Mean

Assume we have a random sample of size n from a NORMAL population with average $\bar{X}$ and standard deviation S. The true mean of the population is μ. Let μ_0 be a fixed number. For the test

$$H_0 : \mu = \mu_0,$$
$$H_a : \mu < \mu_0,$$

the P-value is

$$P = P\left(t(n-1) < \frac{\bar{X} - \mu_0}{S/\sqrt{n}}\right)$$

where $t(n-1)$ is a Student distribution with $n-1$ degrees of freedom. For the test

$$H_0 : \mu = \mu_0,$$
$$H_a : \mu > \mu_0,$$

the P-value is

$$P = P\left(t(n-1) > \frac{\bar{X} - \mu_0}{S/\sqrt{n}}\right).$$

Matched pairs

Assume we want to test the effect of a course on students. We test the students before and after a course to test the effectiveness of the course. We should not analyse such data as two samples. Our two samples techniques work for two *independent* samples. We do not have independence here since we are testing the same individuals in the two samples. We have matched pairs instead. In such a case we should analyse the difference between the two tests for each individual. We may apply a one sample technique to the differences. Next we treat such an example.

Example 3. Ten students are given two similar tests before and after a course. Here are their grades

Before	71	72	85	90	55	61	76	78	79	85
After	73	75	89	92	50	68	82	81	86	80

We start by computing the gain per student. We get 2, 3, 4, 2, −5, 7, 6, 3, 7 and −5. We get an average gain of 2.4 and a sample standard deviation of 4.3. Let μ be the true gain after the course. We would like to test

$$H_0 : \mu = 0,$$
$$H_a : \mu > 0.$$

Assuming that the gains are normally distributed, we use a Student test. The P value is

$$P = P\left(t(9) > \frac{\bar{X}}{S/\sqrt{10}}\right) = P(t(9) > 1.75).$$

A Student table yields that the P value is strictly between 0.05 and 0.1. At the 5% level we cannot reject the null hypothesis. That is, the chance that there is no gain after the course is larger than 5%.

Checking normality

How do we decide whether the assumption of normality is reasonable? This is what the next example is about.

Example 4. Consider the following data: 10.8, 9.6, 10.2, 9.8, 6.9, 8.7, 12.2, 10.4, 11.7, 9.0, 7.4, 13.2, 10.9, 9.5, 11.0, 6.9, 12.9, 6.2, 9.2, 16.9.

Could these numbers come from a normal population? We first draw a histogram.

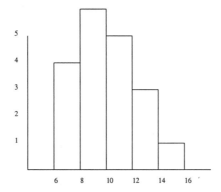

This histogram is nearly bell shaped and symmetric. However, interpreting a histogram is subjective. We may quantify this interpretation by using the fact that 68% of normal observations should be within one standard deviation of the mean and 95% of the observations should be within two standard deviations of the mean. See Exercise 2.

However, there is a more precise way to assess normality which is provided by a normal quantile plot. We have a sample of 20 and the smallest observation is 6.2. So 6.2 is the 1/20 or 0.05 quantile of the data. The 0.05 quantile for a standard normal distribution is the number with an area of 0.05 to its left. So the 0.05 quantile for a standard normal is −1.64. The first point on the normal quantile plot is (6.2, −1.64). The second smallest observation is 6.9. So this is the 2/20 or 0.1 quantile of the data. The 0.1 quantile of a standard normal distribution is -1.28. The second point of our normal quantile plot is $(6.9, -1.28)$ and so on. Below is the normal quantile plot for our data.

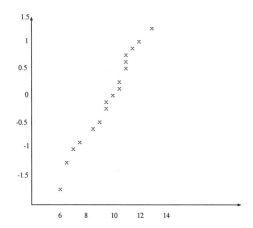

If the distribution of the observations is normally distributed, then the normal quantile plot is close to a line. This is so because if X is normally distributed with mean μ and standard deviation σ, then $\frac{X-\mu}{\sigma}$ is a standard normal distribution. So there is a linear relation between the quantiles of any two normally distributed random variables.

In this particular example one sees that the points are reasonably aligned and we may conclude that our observations come from an approximately normal population.

The sign test

If the population is clearly not normal and the sample is too small to use the CLT, we still have the sign test at our disposal. This is a test that may be performed without assuming that the random variables follow a normal distribution. We still need to have a sample of n i.i.d. random variables but nothing will be assumed about the distribution of these random variables. We will explain the test in an example.

Example 5. Assume that the heights in cm of eleven 6-year old boys are the following: 80,93,85,87,79,85,85,86,89,90 and 91. We would like to test the following claim: the median height of a 6-year old boy in this population is at least 84 cm. So our test is

$$H_0 : m = 84,$$
$$H_a : m > 84.$$

where m is the true median of the population. If the median of the distribution of the continuous random variable X is m, then by definition of m we have

$$P(X > m) = P(X \leq m) = 1/2.$$

Let B be the number of observations above 84. Under the null hypothesis $m = 84$, B is a binomial random variable with parameters $n = 11$ and $p = 1/2$ (since there is the same chance for an observation to fall below or above 84). The sign test is based on the random variable B. We should reject the null hypothesis if B is too large. We compute the P-value for the sign test

$$P = P(B \geq 9 | m = 84) = \binom{11}{9}(1/2)^{11} + \binom{11}{10}(1/2)^{11} + \binom{11}{11}(1/2)^{11}.$$

We get a P-value of 0.03. Thus, at the 5% level we reject the null hypothesis: there is statistical evidence that the median height in this population is at least 84 cm.

Example 6. A diet is tested on eight people. Here are the weights before and after the diet

Before	181	178	205	195	202	176	180	177
After	175	171	196	192	190	168	176	171

Test the claim that the median weight loss for this diet is larger than 5 pounds. Let m be the true median weight loss. We want to test

$$H_0 : m = 5,$$
$$H_a : m > 5.$$

We first compute the weight losses: 6,7,9,3,12,8,4,6. Let B be the number of weight losses larger than 5 pounds. Under $m = 5$, B follows a binomial with parameters $n = 8$ and $p = 1/2$. Thus, the P-value is

$$P = P(B \geq 6 | m = 5) = \binom{8}{6}(1/2)^8 + \binom{8}{7}(1/2)^8 + \binom{8}{8}(1/2)^8 = 0.14.$$

At the level 5% or 10% there is not enough evidence to reject the null hypothesis. That is, there is not enough evidence to claim that the median weight loss of the diet is at least 5 pounds.

Remark. Note that the sign test is a test on the median not the mean. Note also that only a little information from the data is used (we only need to know how many observations are above a certain threshold; we do not need to know the exact values). The sign test is less powerful than a Student test in the following sense. For a given type I error (rejecting the null hypothesis when true) the type II error (rejecting the alternative hypothesis when true) is larger for the sign test than for the Student test.

Exercises

1. (a) What is $P(t(3) > 2)$?
   (b) Compare (a) with $P(Z > 2)$.

2. Consider the observations of Example 4.

(a) What percentage of the observations are within 1 standard deviation of the mean?

(b) What percentage of the observations are within 2 standard deviations of the mean?

(c) Is it reasonable to assume that these observations come from a normal population?

3. Some components in the blood tend to vary normally over time for each individual. Assume that the following levels for a given component were measured on a single patient: 5.5, 5.2, 4.5, 4.9, 5.6 and 6.3.

(a) Test the claim that the mean level for this patient is above 4.7.

(b) Find a confidence interval with 0.95 confidence for the mean level of this patient.

4. Assume that a group of 10 eighth graders taken at random averaged 85 on a test with a standard deviation of 7.

(a) Is there evidence that the true mean grade for this population is above 80?

(b) Find a confidence interval for the true mean grade.

(c) What assumptions did you make to answer (a) and (b)?

5. Eight students were given a placement test and after a week of classes were given again a placement test at the same level. Here are their scores.

Before	71	78	80	90	55	65	76	77
After	75	71	89	92	61	68	80	81

(a) Test whether the scores improved after one week by performing a Student test.

(b) Test whether the scores improved after one week by performing a sign test.

6. In an agricultural field trial, researchers tested two varieties of tomatoes in ten plots. In eight of the plots variety A yielded more than variety B. Is this enough evidence to say that variety A yields more than variety B?

7. A diet was tested on nine people. Here are their weights before and after the diet.

Before	171	178	180	190	165	165	176	177	182
After	175	171	182	161	168	156	165	171	175

(a) Test whether the diet causes a loss of at least 5 pounds by performing a Student test.

(b) Check whether it is reasonable to assume normality in (a).

(c) Perform a sign test for the hypothesis in (a).

8. A test given to 12 male students has an average of 75 and standard deviation of 11. The same test given to 10 female students has an average of 81 with a standard deviation of 8.

(a) Is there evidence that the female students outperform the male students?

(b) Find a confidence interval for the difference of the true means.

Assuming that the two samples are random, independent and come from normal populations you may use that

$$\frac{\bar{X}_1 - \bar{X}_2 - (\mu_1 - \mu_2)}{\sqrt{S_1^2/n_1 + S_2^2/n_2}}$$

follows approximately a student distribution with $n - 1$ degrees of freedom where n is the smallest of n_1 and n_2.

5.4   Chi-Square Tests

In this section we will see two Chi-Square tests. We will first test whether two variables are independent. Our second test will check whether given observations fit a theoretical model. We start by introducing the Chi-Square distributions.

Chi-Square Distributions

The random variable X is said to have a Chi-square distribution with n degrees of freedom if it is a continuous random variable with density

$$f(x) = \frac{1}{2^{n/2}\Gamma(n/2)} x^{n/2-1} e^{-x/2}, \qquad x > 0$$

where

$$\Gamma(a) = \int_0^\infty x^{a-1} e^{-x} dx \text{ for } a > 0.$$

The notation for a Chi-square random variable with n degrees of freedom is $\chi^2(n)$.

The function Γ above may seem strange but it is very important and appears in a number of applications. An integration by parts shows that

$$\Gamma(a + 1) = a\Gamma(a).$$

In particular, for a positive integer n we get that

$$\Gamma(n) = (n - 1)!$$

We now sketch the graphs of three densities of Chi-square random variables.

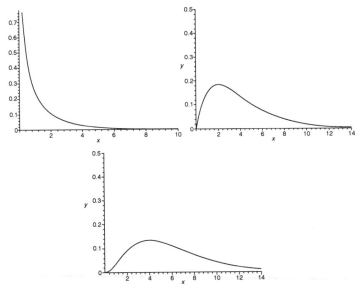

From left to right we have the Chi-square distributions with 1, 4 and 6 degrees of freedom, respectively.

Testing independence

Example 1. Is there a relation between level of education and smoking? Assume that a random sample of 200 was taken with the following results.

	Smoker	Non-Smoker
Education		
8 years or less	9	38
12 years	21	80
16 years	5	47

In the test we are going to perform, the null hypothesis will be that there is no association between the two variables. That is, H_0 will be the hypothesis that education level and smoking are independent. The alternative hypothesis is that there is an association between the two variables. In order to make a decision we will compare the counts in our sample to the expected counts under the null hypothesis. We now explain how to compute the expected counts under the null hypothesis. There are nine people with eight or fewer years of education and who smoke. The probability that someone in the sample has eight or fewer years of education is

$$\frac{9+38}{200} = \frac{47}{200}.$$

The probability that someone in the sample is a smoker is

$$\frac{9 + 21 + 5}{200} = \frac{35}{200}.$$

Thus, under the assumption that level of education and smoking are independent, we get that the probability that someone taken at random in the sample has eight or fewer years of education and who smoke is

$$\frac{47}{200} \times \frac{35}{200}.$$

The expected number of people who have eight or fewer years of education and smoke is therefore

$$200 \times \frac{47}{200} \times \frac{35}{200} = \frac{47 \times 35}{200}.$$

More generally we have the following.

Expected Count under the Independence Assumption

$$\text{Expected count} = \frac{\text{row total} \times \text{column total}}{\text{sample size}}.$$

We now go back to the data of Example 1 and compute the expected counts for all the cells.

	Expected Counts	
	Smoker	**Non-Smoker**
Education		
8 years or less	8.225	38.775
12 years	17.675	83.325
16 years	9.1	42.9

Testing Independence

Assume that we want to test whether two variables are related. The null hypothesis is H_0 : The two variables are independent. We use the statistic

$$X^2 = \sum \frac{(\text{observed} - \text{expected})^2}{\text{expected}}.$$

The random variable X^2 follows approximately a $\chi^2((r-1)(c-1))$ distribution where r and c are the number of rows and columns respectively. Therefore the P value for this test is given by

$$P(\chi^2((r-1)(c-1)) > X^2).$$

The approximation of X^2 by a Chi-square distribution gets better as the sample size increases and is more reliable if every expected cell has a count of 5 or more.

We now go back to the data of Example 1 to perform the test. We compute X^2.

$$
\begin{aligned}
X^2 &= \frac{(9-8.225)^2}{8.225} + \frac{(38-38.775)^2}{38.775} + \frac{(21-17.675)^2}{17.675} \\
&+ \frac{(80-83.325)^2}{83.325} + \frac{(5-9.1)^2}{9.1} + \frac{(47-42.9)^2}{42.9} = 3.09.
\end{aligned}
$$

We have three categories for education so $r = 3$ and two categories for smoking so $c = 2$. Thus, $(r-1)(c-1) = 2$ and X^2 follows approximately a $\chi^2(2)$. The P value for Example 1 is

$$P = P(\chi^2(2) > 3.09).$$

According to the Chi-square table the P value is larger than 0.1. At the 5% level we do not reject H_0. It does not appear that there is an association between education level and smoking.

Goodness-of-fit test

We now turn to another important test: goodness-of-fit. We start with an example.

Example 2. Consider the following 25 observations: 0, 3, 1, 0, 1, 1, 1, 3, 4, 3, 2, 0, 2, 0, 0, 0, 4, 2, 3, 4, 1, 6, 1, 4, 1. Could these observations come from a Poisson distribution?

Recall that a Poisson distribution depends only on one parameter: its mean. We use the sample average to estimate the mean. We get

$$\bar{X} = \frac{47}{25} = 1.88.$$

Let N be a Poisson random variable with mean 1.88. We have that

$$P(N = 0) = e^{-1.88} = 0.15$$

and therefore the expected number of 0s in 25 observations is $25 \times e^{-1.88} = 3.81$. Likewise we have that

$$P(N = 1) = 1.88e^{-1.88} = 0.29$$

and the expected number of 1s in 25 observations is 7.17. We also get that the expected number of 2s is 6.74 and the expected number of 3s is 4.22. The probability that N is 4 or more is

$$P(N \geq 4) = 0.12.$$

Thus, the expected number of observations larger than 4 is 3. We summarize these computations in the table below.

	0	1	2	3	4 or more
Observed	6	7	3	4	5
Expected	3.81	7.17	6.74	4.22	3

The test we are going to perform compares the expected and observed counts in the following way.

Goodness-of-fit

We want to test whether some observations are consistent with a certain distribution F (F may be for instance a Poisson distribution or a normal distribution). The null hypothesis is H_0 : The observations follow a distribution F. We use the statistic

$$X^2 = \sum \frac{(\text{observed} - \text{expected})^2}{\text{expected}}.$$

The random variable X^2 follows approximately a $\chi^2(r - 1 - d)$ distribution where r is the number of categories of observations and d is the number of parameters that must be estimated for the distribution F. Therefore the P value for this test is given by

$$P(\chi^2(r - 1 - d) > X^2).$$

We use the preceding rule on Example 2. In that case the categories are 0, 1, 2, 3 and 4 or more. So $r = 5$. The Poisson distribution depends on one parameter (its mean) therefore $d = 1$. We now compute X^2.

$$X^2 = \frac{(6-3.81)^2}{3.81} + \frac{(7-7.17)^2}{7.17} + \frac{(3-6.74)^2}{6.74}$$
$$+ \frac{(4-4.22)^2}{4.22} + \frac{(5-3)^2}{3} = 4.68$$

We know that X^2 follows approximately a Chi-square distribution with $r - 1 - d = 5 - 1 - 1 =$ three degrees of freedom so the P value is

$$P = P(\chi^2(3) > 4.68).$$

Since the P-value is larger than 0.1 we do not reject the null hypothesis. There is no evidence that these observations do not follow a Poisson distribution.

Example 3. The observations of Example 2 were in fact generated as Poisson observations with mean 2 by a computer random generator. We now test whether these observations are consistent with a Poisson distribution with mean 2. That is, our null hypothesis is now H_0: The observations follow a Poisson distribution with mean 2. The only difference with Example 1 is that now we do not need to estimate the mean of the Poisson distribution. In particular, for this example $d = 0$. We compute the expected counts.

	0	1	2	3	4 or more
Observed	6	7	3	4	5
Expected	3.38	6.77	6.77	4.51	3.75

This time $X^2 = 4.62$. We have that $r - d - 1 = 5 - 0 - 1 = 4$.

$$P = P(\chi^2(4) > 4.62) > 0.1.$$

We do not reject the null hypothesis. There is no evidence that these observations do not follow a Poisson distribution with mean 2.

The following example deals with continuous distributions.

Example 4. Are the following observations consistent with a normal distribution?

66, 64, 59, 65, 81, 82, 64, 60, 78, 62
65, 67, 67, 80, 63, 61, 62, 83, 78, 65
66, 58, 74, 65, 80

The sample average is 69 and the sample standard deviation is 8 (we are rounding to the closest 1 to simplify the subsequent computations). We will now try to fit the observations to a normal distribution with mean 69 and standard deviation 8.

We first pick the number of categories, keeping in mind that the Chi-square approximation is best when there are at least 5 expected counts per cell. We pick 5 categories. Using the standard normal table we find the 20th, 40th, 60th and 80th percentiles. For instance, we read in the standard normal table that

$$P(Z < -0.84) = 0.2$$

and so the 20th percentile of a standard normal distribution is -0.84. Likewise we find the four percentiles in increasing order

$-0.84, -0.25, 0.25, 0.84$

Recall that if X is a normal random variable with mean 69 and standard deviation 8, then

$$\frac{X - 69}{8}$$

is a standard normal random variable. So, for instance, the 20th percentile of a normal random variable with mean 69 and standard deviation 8 is

$$69 + 8(-0.84) = 62.28.$$

Likewise the 40th, 60th, and 80th percentiles of a normal random variable with mean 69 and standard deviation 8 are $67, 71, 75.72$. We round these percentiles to the nearest one. We now compare the observed and expected counts.

Category	$(-\infty, 62]$	$(62, 67]$	$(67, 71)$	$[71, 76)$	$[76, \infty)$
Observed	6	11	0	1	7
Expected	5	5	5	5	5

We compute the statistic

$$X^2 = \sum \frac{(\text{observed} - \text{expected})^2}{\text{expected}} = 16.4.$$

We had to estimate two parameters (μ and σ) so $d = 2$ and X^2 is approximately a Chi-square random variable with $r - d - 1 = 5 - 2 - 1 = 2$ degrees of freedom. We get the P value

$$P(\chi^2(2) > 16.4) < 0.01.$$

So we reject the null hypothesis. These observations are not consistent with a normal distribution.

Exercises

Problems 1 through 5 use data from the American Mathematical Society regarding new doctorates in mathematics (*Notices of the AMS* January 1998). Types I, II and III are groups of mathematics departments as ranked by the AMS.

1. The following table gives the number of new PhD's in mathematics according to their area of concentration and the type of department that granted their degree.

Area Institution	Algebra	Geometry	Probability and Stat.
I	21	28	9
II	10	7	4
III	3	1	3

Is it true that certain types of institutions graduate more students in one area than in others?

2. The table below breaks down the number of graduates in 1997 according to their gender and area of concentration.

Area Gender	Algebra	Geometry	Probability and Stat.
Male	123	118	194
Female	37	23	98

Is the distribution of area of concentrations for female students different from the distribution for male students?

3. The following table gives the numbers of employed new graduates according to the type of their granting institution and the type of employer.

Granting Inst. Employer	I public	I private	II
I public	35	19	6
I private	13	25	2
II	14	7	16

Does the type of employer depend on the type of granting institution?

4. The next table breaks down the number of new graduates according to gender and granting institution.

Granting Inst. Gender	I public	I private	II	III
Male	239	157	175	96
Female	58	30	63	36

Is the distribution of granting institutions for female students different from the distribution for male students?

5. The table below breaks down the number of employed new graduates per type of employer and citizenship.

Citizenship	US	Non-US
Employer		
PhD dept.	100	111
Non-PhD dept.	177	59
Non-academic	104	160

Does the type of employer depend on citizenship?

6. Test whether the following observations are consistent with a Poisson distribution: 1, 4, 2, 7, 4, 3, 0, 2, 5, 2, 3, 2, 1, 5, 5, 0, 3, 2, 2, 2, 2, 1, 4, 1, 2, 4.

7. Test whether the following observations are consistent with a standard normal distribution:

1.70, 0.11, 0.14, 0.81, 2.19
$-1.56, -0.67, 0.89, -1.24, 0.26$
$-0.05, 0.72, 0.29, -1.09, -0.43$
$-2.23, -1.68, 0.23, 1.17, -0.87$
$-0.28, 1.11, -0.43, -0.16, -0.07$

8. Test whether the following observations are consistent with a uniform distribution on [0,100].

99,53,18,21,20,53,58,4,32,51,
24,51,62,98,2,48,97,64,61,18,
25,57,92,72,95

9. Let T be an exponential random variable with mean 2. That is, the density of T is $f(t) = \frac{1}{2}e^{\frac{-t}{2}}$ for $t \geq 0$. Find the 20th percentile of T.

10. Test whether the following observations are consistent with an exponential distribution.

13,7,14,10,12,8,8,8,10,9,
8,10,5,14,13,7,11,11,10,8,
10,10,13,9,10

11. (a) Show that the function Γ:

$$\Gamma(a) = \int_0^\infty x^{a-1}e^{-x}dx$$

is defined on $(0, \infty)$.
  (b) Show that $\Gamma(a + 1) = a\Gamma(a)$.
  (c) Show that $\Gamma(n) = (n - 1)!$ for any positive integer n.

6

Linear Regression

6.1 Fitting a Line to Data

We start by an example.

Example 1. We consider state taxes and state debts, per capita, for 10 American states (data from The World Almanac and Book of Facts 1998).

x = **Debt**	884	720	798	1526	899	4719	4916	1085	781	4377
y = Taxes	1194	1475	1365	1686	1209	2282	2224	1311	1317	2422

We would like to see whether there is a linear relationship between per capita taxes and per capita debt. We start by plotting the points. Let x denote the state debt per capita and y denote the tax per capita. We get

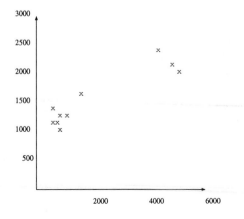

It appears that there is an approximate linear relation between x and y. How can we find a line that fits the data? The most used criterion is the least-squares criterion.

The Least-Squares Regression Line

Assume that we have n observations (x_i, y_i) and we want the line that best fits these observations. More precisely, we would like to predict y from x by using a line. The line $bx + a$ is said to be the least-squares regression line of y on x if

$$\sum_{i=1}^{n} (y_i - (bx_i + a))^2$$

is minimum for $a = \hat{a}$ and $b = \hat{b}$. The values of $\hat{a}$ and $\hat{b}$ are given by

$$\hat{b} = \frac{n \sum_{i=1}^{n} x_i y_i - \sum_{i=1}^{n} x_i \sum_{i=1}^{n} y_i}{n \sum_{i=1}^{n} x_i^2 - (\sum_{i=1}^{n} x_i)^2}$$

and

$$\hat{a} = \bar{y} - \hat{b}\bar{x}.$$

Remarks.

1. Note that $\sum_{i=1}^{n} (y_i - (bx_i + a))^2$ represents the total error we make when we replace y_i by $bx_i + a$. This is why we want a and b to minimize this sum. Other choices are possible for the error, such as $\sum_{i=1}^{n} |y_i - (bx_i + a)|$. However, the advantage of the sum of squares is that we can get explicit expressions for $\hat{a}$ and $\hat{b}$. Explicit expressions are not available if the error is taken to be the sum of absolute values.

2. Note also that the variables x and y do not play symmetric roles here. We are trying to predict y from x. If we want to predict x from y, then we should minimize $\sum_{i=1}^{n} (x_i - (by_i + a))^2$. We would get the regression line of x on y and the reader may check that this is a different line from the regression line of y on x.

3. Finally, note that the relation $\hat{a} = \bar{y} - \hat{b}\bar{x}$ shows that the regression line passes through the point of averages $(\bar{x}, \bar{y})$.

We now go back to the data of Example 1 and compute $\hat{a}$ and $\hat{b}$. We write the intermediate computations in a table.

	x_i	y_i	x_i^2	y_i^2	$x_i y_i$
	884	1194	781456	1425636	1055496
	720	1475	518400	2175625	1062000
	798	1365	636804	1863225	1089270
	1526	1686	2328676	2842596	2572836
	899	1209	808201	1461681	1086891
	4719	2282	22268961	5207524	10768758
	4916	2224	24167056	4946176	10933184
	1085	1311	1177225	1718721	1422435
	781	1317	609961	1734489	1028577
	4377	2422	19158129	5866084	10601094
Sums:	20705	16485	72454869	29241757	41620541

We get that

$$\hat{b} = \frac{10(41620541) - (20705)(16485)}{10(72454869) - (20705)^2} \sim 0.25$$

and that

$$\hat{a} = \bar{y} - \hat{a}\bar{x} = 1648.5 - (0.25)2070.5 \sim 1131,$$

so that the equation of the regression line is

$$y = 0.25x + 1131.$$

One can see below that the regression line fits well the scatter plot.

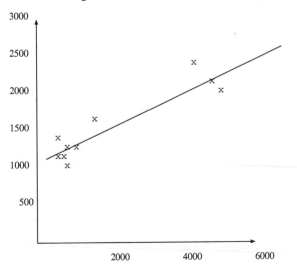

The regression line's main use is to make predictions. For instance, if the debt per capita is $6000 in a state, then we plug $x = 6000$ in the regression line and get that the predicted tax y is

$$y = 0.25 \times 6000 + 1131 = 2631.$$

Example 2. We consider the population, in millions, of the United States from 1790 to 1900.

Year	Population	Year	Population
1790	3.9	1850	23.2
1800	5.3	1860	31.4
1810	7.2	1870	39.8
1820	9.6	1880	50.2
1830	12.9	1890	62.9
1840	17.1	1900	76.0

The scatter plot below shows that there is a relation between population and year but that this relation is not linear.

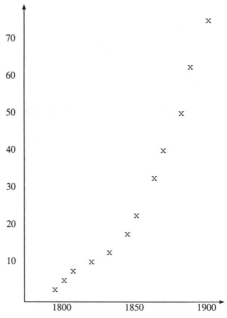

It appears that the population has grown exponentially over the period of time we consider. Let x be the year and y the population; if y is an exponential function of x as

$$y = ce^{dx},$$

then by taking logarithms on both sides we get

$$\ln(y) = dx + \ln(c).$$

Therefore, we may test our hypothesis that the population grew exponentially fast during the period 1790–1900 by trying to find a linear relation between the logarithm of the population and the year. Our transformed data is

Year	ln(Population)	Year	ln(Population)
1790	1.36	1850	3.14
1800	1.67	1860	3.45
1810	1.97	1870	3.68
1820	2.26	1880	3.92
1830	2.56	1890	4.14
1840	2.84	1900	4.33

The regression line fits the data remarkably well. The equation of the regression line is

$$\ln(y) = 0.0275x - 47.77.$$

We plug $x = 1872$ in the preceding equation to predict the population in 1872. We get $\ln(y) = 3.71$ and therefore $y = 40.8$. Thus, the model predicts that the population in 1872 was 40.8 millions in the United States. This figure is in good agreement with the actual figure. We plug $x = 2,000$ and we get $y = 1,380.2$. The prediction of the model is that the population of the United States in 2,000 will be 1 billion and 380 millions people. This is in gross disagreement with the actual figure (around 260 millions). We used the data from 1790 to 1900 to get this regression line. The year 2,000 is well off this range. This example illustrates the fact that, as we get away from the range of x for which the regression was made, the predictions may become very unreliable.

Sample correlation

When one looks for a relation between x and y the starting point should always be a scatter plot. However, the relation between x and y is rarely obvious and it is interesting to have a measure of the strength of the linear relation between x and y. This is where correlation comes into play.

Sample Correlation

Assume that we have n observations of the variables x and y that we denote by (x_i, y_i). The sample correlation r for this data is

$$r = \frac{1}{n-1} \sum_{i=1}^{n} \left(\frac{x_i - \bar{x}}{s_x} \right) \left(\frac{y_i - \bar{y}}{s_y} \right)$$

where $\bar{x} = \frac{1}{n} \sum_{i=1}^{n} x_i$ and $s_x = \frac{1}{n-1} \sum_{i=1}^{n} (x_i - \bar{x})^2$. The coefficient r is always in the interval $[-1, 1]$ and measures the strength of the LINEAR relation between x and y. If $|r|$ is close to 1, there is a strong linear relation between x and y. If $|r|$ is close to 0, then there is no linear relation between x and y. A computational formula for r is

$$r = \frac{\sum_{i=1}^{n} x_i y_i - \frac{1}{n} \sum_{i=1}^{n} x_i \sum_{i=1}^{n} y_i}{(n-1)s_x s_y}.$$

Next we examine three typical situations.

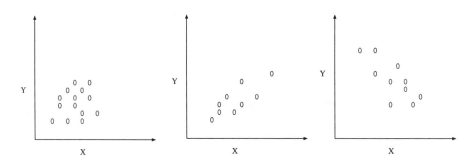

For the scatter plot on the left there is no linear relation between x and y. In this case the correlation coefficient r will be close to 0. For the scatter plot in the middle there is a positive linear relation between x and y and so r will be close to 1. Finally, for the scatter plot on the right there is a negative relation between x and y, and so r will be close to -1. It is possible to show that $r = 1$ or -1 if and only if all the points are aligned.

Example 3. We compute the sample correlation for the data of Example 1. We use the computational formula for s_x and s_y,

$$s_x^2 = \frac{1}{n-1} \left(\sum_{i=1}^{n} x_i^2 - n(\bar{x})^2 \right) = \frac{1}{9}(72454869 - 10 \times (2070.5)^2) = 3287241.$$

Similarly,

$$s_y^2 = \frac{1}{9}(29241757 - 10 \times (1648.5)^2) = 229582.$$

Therefore we get that $s_x = 1813$ and $s_y = 479$. Thus, the correlation coefficient

$$r = \frac{\sum_{i=1}^n x_i y_i - \frac{1}{n}\sum_{i=1}^n x_i \sum_{i=1}^n y_i}{(n-1)s_x s_y} = \frac{41620541 - \frac{1}{10}20705 \times 16485}{9(1813)(479)} \sim 0.96.$$

This computation shows a very strong positive linear relation between tax and debt.

We now state an interesting relation between the correlation coefficient and the equation of the regression line.

Correlation and Regression

Let r be the sample correlation for variables x and y. Let $\hat{b}x + \hat{a}$ be the equation of the regression line of y on x; then

$$\hat{b} = r\frac{s_y}{s_x}$$

and

$$\hat{a} = \bar{y} - \hat{b}\bar{x}.$$

Example 4. We compute the equation of the regression line of y on x for the data in Example 1 by using the formulas above. First,

$$\hat{b} = r\frac{s_y}{s_x} = 0.96\frac{479}{1813} = 0.25.$$

For $\hat{a}$ we use

$$\hat{a} = \bar{y} - \hat{b}\bar{x} = 1648.5 - (0.25)(2070.5) = 1131.$$

Exercises

1. Here is the population data for the United States from 1900 to 1990.

Year	Population	Year	Population
1900	76.0	1950	151.3
1910	92.0	1960	179.3
1920	105.7	1970	203.3
1930	122.8	1980	226.5
1940	131.7	1990	248.7

(a) Make a scatter plot.

(b) Did the population increase exponentially fast during this period?

(c) Do a regression explaining the log of the population as a function of the year.

(d) Does the model in (c) look adequate?

(e) Predict the population in 2,000 by using this model.

2. Use the data of Example 1 to do a regression where the roles of x and y are inverted. That is, take y to be the debt per capita and x the tax per capita.
(a) Find the equation of the regression line.

(b) Is this the same line as the one we found in Example 1?

3. As the scatter plot from Example 1 shows, the data we have analyzed so far has two clusters. One with small tax debt and the other one with high tax debt. We now add four points with intermediate tax debt, the first coordinate represents the debt, the second one the corresponding tax: (3066,1713), (3775, 1891), (2282, 952) and (2851,1370). So the modified data of Example 1 has now 14 points.
(a) Compute the correlation coefficient debt/tax for the modified data.

(b) Compute the equation of the new regression line.

(c) Compare the fit of the regression line in this problem to the fit of the regression line in Example 1.

4. In recent years there has been a significant increase of tuberculosis in the US. We will examine whether there seems to be a linear relation between the number of cases of AIDS and the number of cases of tuberculosis. In the following table we write the rate per 100,000 population for nine states in 1998 (the data are from the Center for Disease Control).

State	Tuberculosis	AIDS
California	11.8	367
Florida	8.7	538
Georgia	8.3	307
Illinois	7.1	189
Maryland	6.3	390
New Jersey	7.9	496
New York	11.0	715
Pennsylvania	3.7	175
Texas	9.2	284

(a) Draw a scatter plot.
(b) Compute the correlation coefficient between AIDS and tuberculosis.

5. Here are the death rates (per 100,000 persons) per age in the US in 1978 (data from US Department of Health).

Age	Death rate	Age	Death rate
42	296.1	67	2463.0
47	471.6	72	3787.4
52	742.4	77	6024.2
57	1115.9	82	8954.0
62	1774.2		

(a) Draw a scatter plot.
(b) Based on (a) should you transform the data in order to have linear relation?
(c) Compute an adequate regression line.

6. The following table gives the per capita health expenditures in the US.

Year	Expenditure	Year	Expenditure
1940	30	1970	367
1950	82	1971	394
1955	105	1972	438
1960	146	1973	478
1965	211	1974	534
		1975	604

(a) Do a scatter plot.
(b) Do a regression of the log of expenditures on the year.

(c) Compute the correlation coefficient for log of expenditures and year.

(d) Does the model in (b) seem adequate?

7. The following table gives the death rates (100,000 persons) for cancer of the respiratory system in the US.

Year	Rate	Year	Rate
1950	14.5	1970	47.6
1955	20.6	1975	56.7
1960	30.5	1977	61.5
1965	36	1978	62.5

(a) Draw a scatter plot.

(b) Find the regression line for the death rate on the year.

(c) What is the correlation coefficient for the rate and year?

6.2   Inference for Regression

The regression line introduced in the preceding section is based on a sample. If we change the sample we will get another regression line. Therefore, it is important to know how much confidence we may have on our regression line. In particular, in this section we will get confidence intervals for the coefficients of the regression line. In order to do so we need an underlying probability model that we now formulate. We will assume that the variable y we wish to explain is random and that the explanatory variable x is deterministic (that is, non-random). We have n observations (x_i, y_i) for $i = 1, 2, \ldots, n$. We assume that

$$y_i = bx_i + a + e_i$$

where the e_i are random variables. We also assume that the e_i are independent and normally distributed with mean 0 and standard deviation σ. Observe that our model has three parameters a, b and σ. Note also that since $bx_i + a$ is a constant and $E(e_i) = 0$ we get

$$E(y_i) = bx_i + a + E(e_i) = bx_i + a.$$

That is, the model assumes that the mean of a variable y_i corresponding to x_i is $bx_i + a$. However, the random variable y_i varies around its mean. The model assumes that e_i, the variation of y_i around its mean, is normally distributed with standard deviation σ. Therefore, the model makes four major assumptions: the mean of y_i is a linear function of x_i, the variables e_i are normally distributed, the standard deviation is the same for all the e_i and the e_i are independent of each other.

To estimate a and b we use the estimators $\hat{a}$ and $\hat{b}$ that we have given in Section 6.1. To predict the y corresponding to x_i we use

$$\hat{y}_i = \hat{b}x_i + \hat{a}.$$

Let

$$\hat{e}_i = y_i - \hat{y}_i$$

represent the error between the observation y_i and the prediction $\hat{y}_i$.

Estimating the Regression Parameters

Assume that

$$y_i = bx_i + a + e_i \text{ for } i = 1 \ldots n$$

and that the e_i are independent and normally distributed with mean 0 and standard deviation σ. Then a and b are estimated by

$$\hat{b} = \frac{n \sum_{i=1}^{n} x_i y_i - \sum_{i=1}^{n} x_i \sum_{i=1}^{n} y_i}{n \sum_{i=1}^{n} x_i^2 - (\sum_{i=1}^{n} x_i)^2}$$

and

$$\hat{a} = \bar{y} - \hat{b}\bar{x}.$$

The parameter σ^2 is estimated by

$$s^2 = \frac{1}{n-2} \sum_{i=1}^{n} (y_i - (\hat{b}x_i + \hat{a}))^2 = \frac{1}{n-2} \sum_{i=1}^{n} \hat{e}_i^2.$$

Note that to get s^2 we divide by $n-2$. This is similar to what we do to compute the sample variance of a random variable (we divide by $n-1$). By dividing by $n-2$ we get an unbiased estimator of σ^2. That is,

$$E(s^2) = \sigma^2.$$

Example 1. We use the data of Example 1 in Section 6.1 to get estimates of the regression parameters. We have already computed the regression line

$$y = 0.25x + 1131.$$

So to get the predicted $\hat{y}_i$ below we compute

$$\hat{y}_i = 0.25x_i + 1131.$$

	x_i	y_i	$\hat{y}_i$	$\hat{e}_i$	$\hat{e}_i^2$
	884	1194	1352	−158	24964
	720	1475	1311	164	26896
	798	1365	1331	−34	1156
	1526	1686	1513	−173	29929
	899	1209	1356	−147	21609
	4719	2282	2311	−29	841
	4916	2224	2360	−136	18496
	1085	1311	1402	−91	8281
	781	1317	1326	−9	81
	4377	2422	2225	197	38809
Sum:					171062

Thus, we get that

$$s^2 = \frac{1}{n-2} \sum_{i=1}^{n} \hat{e}_i^2 = \frac{1}{8} 171062 = 21383.$$

Therefore, $s = 146$ is an estimate of σ. We will now use the estimators $\hat{a}$, $\hat{b}$ and s to get confidence intervals for a and b. We first need information about the distributions of $\hat{a}$ and $\hat{b}$.

Distribution of $\hat{a}$ and $\hat{b}$

Consider the model

$$y_i = bx_i + a + e_i$$

with the assumptions that the e_i are independent, normally distributed with mean 0 and variance σ^2. Assume that we have n observations. Then,

$$\frac{\hat{a} - a}{s_{\hat{a}}} \quad \text{and} \quad \frac{\hat{b} - b}{s_{\hat{b}}}$$

follow a Student distribution with $n - 2$ degrees of freedom where,

$$s_{\hat{a}} = s \sqrt{\frac{1}{n} + \frac{\bar{x}^2}{\sum_{i=1}^{n}(x_i - \bar{x})^2}}$$

and

$$s_{\hat{b}} = \frac{s}{\sqrt{\sum_{i=1}^{n}(x_i - \bar{x})^2}}.$$

Note that the above holds only if we assume that the e_i are independent, *normally* distributed with mean 0 and the *same* σ. The statistical analysis we will do below is only valid under these assumptions. We will discuss below how to check these assumptions.

The crucial step in any regression analysis is to test whether or not $b = 0$. If we cannot reject the null hypothesis $b = 0$, then the conclusion should be that there is no statistical evidence that there is a linear relation between the variables y and x. In other words our model is not adequate for the problem and we have to look for another model.

Example 2. We are going to test whether $b = 0$ for the data of Example 1. In order to perform the test we need to compute

$$\sum_{i=1}^{n}(x_i - \bar{x})^2 = \sum_{i=1}^{n} x_i^2 - 2\bar{x} \sum_{i=1}^{n} x_i + n\bar{x}^2 = \sum_{i=1}^{n} x_i^2 - \frac{1}{n}(\sum_{i=1}^{n} x_i)^2.$$

We now may use the table in Example 1, Section 6.1 to get

$$\sum_{i=1}^{n}(x_i - \bar{x})^2 = 72454869 - \frac{1}{10}(20705)^2 = 29585166.5.$$

From Example 1, we know that $s = 146$. Thus,

$$s_{\hat{b}} = \frac{s}{\sqrt{\sum_{i=1}^{n}(x_i - \bar{x})^2}} = \frac{146}{\sqrt{29585166.5}} = 0.03.$$

We have already computed $\hat{b} = 0.25$. Since $\hat{b} > 0$, the test we want to perform is

$$H_0 : b = 0,$$
$$H_a : b > 0.$$

We now compute the P value for this test.

$$P(\hat{b} > 0.25 | b = 0) = P\left(t(n-2) > \frac{0.25}{0.03}\right) = P(t(8) > 8).$$

According to the Student table this P value is extremely small and we reject the null hypothesis with very high confidence. That is, there is strong statistical evidence that there is a positive relation between x (state debt) and y (state tax).

Example 3. Since we have rejected the hypothesis $b = 0$ we should find a confidence interval for b. For a confidence interval with 0.95 confidence we want c such that

$$P(|\hat{b} - b| < c) = 0.95.$$

Therefore,

$$P(|\hat{b} - b| < c) = P\left(\frac{|\hat{b} - b|}{s_{\hat{b}}} < \frac{c}{s_{\hat{b}}}\right) = P(|t(n-2)| < \frac{c}{s_{\hat{b}}}) = 0.95.$$

We use the Student table to get

$$\frac{c}{s_{\hat{b}}} = 2.3.$$

So

$$c = 2.3 \times 0.03 = 0.07.$$

Therefore, a confidence interval at the 95% level for b is (0.18,0.32).

We now turn our attention to confidence intervals for the predicted and mean values of y. For a given value x_0 of the variable x, there are two possible interpretations for $\hat{b}x_0 + \hat{a}$. It could be an estimate of the mean value of y corresponding to x_0 or it could be a prediction for a y corresponding to x_0. As we are going to see below there is more variation in predicting an individual y than in predicting its mean.

Predicting y

For a given x_0 there are two possible interpretations for $\hat{b}x_0 + \hat{a}$. It may represent an estimate of the $E(y)$ for a y corresponding to x_0. Or it may represent a prediction for an individual y corresponding to x_0. We have that,

$$\frac{(\hat{b}x_0 + \hat{a}) - (bx_0 + a)}{s_y}$$

follows a Student distribution with $n-2$ degrees of freedom. The value of s_y depends on the interpretation given to $\hat{b}x_0 + \hat{a}$: if $\hat{b}x_0 + \hat{a}$ estimates the expected value of y, then

$$s_y = s\sqrt{\frac{1}{n} + \frac{(x_0 - \bar{x})^2}{\sum_{i=1}^{n}(x_i - \bar{x})^2}}.$$

If $\hat{b}x_0 + \hat{a}$ is used to predict an individual value of y, then

$$s_y = s\sqrt{1 + \frac{1}{n} + \frac{(x_0 - \bar{x})^2}{\sum_{i=1}^{n}(x_i - \bar{x})^2}}.$$

Note that the standard error for prediction is larger than the standard error for estimating the mean.

Example 4. We will now compute a confidence interval for the mean tax per capita corresponding to debt per capita of $ 3,000. We use the data of Example 1. We have $x_0 = 3,000$. In Example 2 we have already computed

$$\sum_{i=1}^{n}(x_i - \bar{x})^2 = 29585166.5.$$

We also have that $\bar{x} = 2070.5$ and that $s = 146$. Therefore,

$$s_y = s\sqrt{\frac{1}{n} + \frac{(x_0 - \bar{x})^2}{\sum_{i=1}^{n}(x_i - \bar{x})^2}} = 146\sqrt{\frac{1}{10} + \frac{(3000 - 2070.5)^2}{29585166.5}} = 52.$$

In this example we have that

$$\hat{b}x_0 + \hat{a} = 1881.$$

Since $\frac{(\hat{b}x_0 + \hat{a}) - (bx_0 + a)}{s_y}$ follows a Student distribution with $n - 2 = 8$ degrees of freedom, we get that a confidence interval with 95% confidence for the mean tax corresponding to a debt of 3,000 per capita is

$$(1881 - 2.3s_y, 1881 + 2.3s_y) = (1761, 2001).$$

Example 5. We now compute a confidence interval for a predicted tax based on a debt of $ 3,000. The only difference with Example 4 is that the standard deviation s_y is now

$$s\sqrt{1 + \frac{1}{n} + \frac{(x_0 - \bar{x})^2}{\sum_{i=1}^{n}(x_i - \bar{x})^2}} = 155.$$

Note that the standard deviation has tripled compared to Example 4. With 95% confidence we get that the predicted tax corresponding to 3,000 debt is in the interval

$$(1881 - 2.3s_y; 1881 + 2.3s_y) = (1515, 2238).$$

Checking the assumptions of the model

We will now check some of the assumptions we made for the model. We start by plotting the residuals for the data of Example 1.

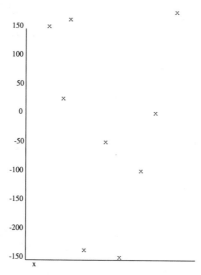

The important thing to check here is that there is no special pattern. For instance, it could be that the residuals increase, decrease in a regular way or have a special clustering. In this case there appears to be no special pattern emerging.

The second important plot is the normal quantile plot for the residuals (see Section 5.3 for more details). We now plot the normal quantile plot for the residuals in Example 1.

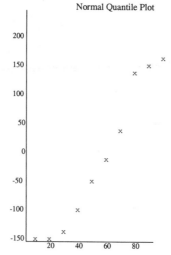

Recall that when the variable is normally distributed the points in this plot are aligned. The pattern here is not too far from a straight line so the assumption of normality seems reasonable in this case. In summary, based on the two plots above we may conclude that the assumptions of the model (normality and independence of the residuals, same σ) are not violated in this example.

Exercises

1. (a) Test whether a is 0 for the data of Example 1.
   (b) Find a confidence interval for a with confidence 0.99.

2. Consider the data about the US population in Exercise 1 in Section 6.1.
   (a) Do a regression of the log of the population on the year.
   (b) Test whether $b = 0$.
   (c) Give a confidence interval for the U.S. population in 2000.

3. Consider the data on death rates from Exercise 5 in Section 6.1.
   (a) Compute the regression line of log of death rate on age.
   (b) Test whether $b = 0$ for the model in (a).
   (c) Plot the residuals of the model in (a).
   (d) Plot the normal probability quantiles for the residuals.
   (e) Based on (c) and (d) would you say that the assumptions of the model hold
in this case?

4. Consider the data in Exercise 6 in Section 6.1.
   (a) Do a regression of the log of expenditures on the year.
   (b) Test the adequacy of the model.
   (c) Test the assumptions of the model.

5. Consider the data of Exercise 4 in Section 6.1.
   (a) Do a regression of the tuberculosis rate on the AIDS rate.
   (b) Test whether there is a linear relation between the two rates.

6. Consider the data of Exercise 7 in Section 6.1 about the death rate for cancer of
the respiratory system.
   (a) Test whether there is a linear relation between death rate and year.
   (b) Give a confidence interval for the death rate of year 1985.

7. The table below gives the number of years a person alive at 65 in a given year is
expected to live.

Year	Life expectancy	Year	Life expectancy
1900	11.9	1975	16.0
1950	13.9	1977	16.3
1960	14.3	1979	16.3
1970	15.2		

   (a) Test whether there is a linear relation between life expectancy and year.
   (b) Give a confidence interval for the life expectancy at age 65 in the year 2000.
   (c) Does the answer in (b) look accurate?

7

Moment Generating Functions and Sums of Independent Random Variables

7.1 Moment Generating Functions

The purpose of this chapter is to introduce moment generating functions (mgf). We have two applications in mind that will be covered in the next section. We will compute the distribution of some sums of independent random variables and we will indicate how moment generating functions may be used to prove the Central Limit Theorem. We start with their definition.

Moment Generating Functions

The moment generating function of a random variable X is defined as

$$M_X(t) = E(e^{tX}).$$

In particular, if X is a discrete random variable, then

$$M_X(t) = \sum_k e^{tk} P(X = k).$$

If X is a continuous random variable and has a density f, then

$$M_X(t) = \int e^{tx} f(x) dx.$$

Note that an mgf is not necessarily defined for all t (because of convergence problems of the series or of the generalized integral). But it is useful even if it is defined only on a small interval. We start by computing some examples.

Example 1. Consider a binomial random variable S with parameters n and p. Compute its mgf.

We have that

$$
\begin{aligned}
M_S(t) &= E(e^{tS}) = \sum_{k=0}^{n} e^{tk} P(S=k) = \sum_{k=0}^{n} e^{tk} \binom{n}{k} p^k (1-p)^{n-k} \\
&= \sum_{k=0}^{n} \binom{n}{k} (e^t p)^k (1-p)^{n-k}.
\end{aligned}
$$

We now use the binomial Theorem

$$
(x+y)^n = \sum_{k=0}^{n} \binom{n}{k} x^k y^{n-k}
$$

with $x = e^t p$ and $y = (1-p)$ to get

$$
M_S(t) = (pe^t + 1 - p)^n \text{ for all } t.
$$

Example 2. Let N be a Poisson random variable with mean λ. We have

$$
M_N(t) = E(e^{tN}) = \sum_{k=0}^{\infty} e^{tk} P(N=k) = \sum_{k=0}^{\infty} e^{tk} e^{-\lambda} \frac{\lambda^k}{k!} = \sum_{k=0}^{\infty} e^{-\lambda} \frac{(e^t \lambda)^k}{k!}.
$$

Recall that

$$
e^x = \exp(x) = \sum_{k=0}^{\infty} \frac{x^k}{k!}.
$$

We use this power series expansion with $x = e^t \lambda$ to get

$$
M_N(t) = e^{-\lambda} \exp(e^t \lambda) = \exp(\lambda(-1 + e^t)) \text{ for all } t.
$$

We now give an example of computation of an mgf for a continuous random variable.

Example 3. Assume X is exponentially distributed with rate λ. Its mgf is

$$
M_X(t) = E(e^{tX}) = \int_0^{\infty} e^{tx} \lambda e^{-\lambda x} dx = \int_0^{\infty} \lambda e^{(t-\lambda)x} dx.
$$

Note that the preceding improper integral is convergent only if $t - \lambda < 0$. In that case we get

$$
M_X(t) = \frac{\lambda}{\lambda - t} \text{ for } t < \lambda.
$$

The moment generating functions get their name from the following property.

Moments of a Random Variable

Let X be a random variable. The expectation $E(X^k)$ is called the kth moment of X. If X has a moment generating function M_X defined on some interval $(-r, r)$ for $r > 0$, then all the moments of X exist and

$$E(X^k) = M_X^{(k)}(0)$$

where $M_X^{(k)}$ designates the kth derivative of M_X.

Example 4. We will use the formula above to compute the moments of the Poisson distribution. Let N be a Poisson random variable with mean λ. Then M_N is defined everywhere and

$$M_N(t) = \exp(\lambda(-1 + e^t)).$$

Note that the first derivative is

$$M_N'(t) = \lambda e^t \exp(\lambda(-1 + e^t)).$$

Letting $t = 0$ in the formula above yields

$$E(X) = M_N'(0) = \lambda.$$

We now compute the second derivative

$$M_N''(t) = \lambda e^t \exp(\lambda(-1 + e^t)) + \lambda^2 e^{2t} \exp(\lambda(-1 + e^t)).$$

Thus,

$$E(X^2) = M_N''(0) = \lambda + \lambda^2.$$

Note that

$$Var(X) = E(X^2) - E(X)^2 = \lambda.$$

Example 5. We now compute the mgf of a standard normal distribution. Let Z be a standard normal distribution. We have

$$M_Z(t) = E(e^{Zt}) = \int_{-\infty}^{\infty} \frac{1}{\sqrt{2\pi}} e^{zt} e^{-z^2/2} dz = \int_{-\infty}^{\infty} \frac{1}{\sqrt{2\pi}} e^{zt - z^2/2} dz.$$

Note that we may 'complete the square' to get

$$zt - z^2/2 = -(z - t)^2/2 + t^2/2.$$

Thus,

$$M_Z(t) = \int_{-\infty}^{\infty} \frac{1}{\sqrt{2\pi}} e^{-(z-t)^2/2 + t^2/2} dz = e^{t^2/2} \int_{-\infty}^{\infty} \frac{1}{\sqrt{2\pi}} e^{-(z-t)^2/2} dz.$$

Note that $g(z) = \frac{1}{\sqrt{2\pi}} e^{-(z-t)^2/2}$ is the density of a normal distribution with mean t and standard deviation 1. Thus,

$$\int_{-\infty}^{\infty} \frac{1}{\sqrt{2\pi}} e^{-(z-t)^2/2} dz = 1$$

and

$$M_Z(t) = e^{t^2/2}.$$

Example 6. We may use Example 5 to compute the moments of a standard normal distribution.

$$M_Z'(t) = te^{t^2/2}.$$

Letting $t = 0$ above we get

$$E(Z) = 0.$$

We have

$$M_Z''(t) = e^{t^2/2} + t^2 e^{t^2/2}.$$

So

$$E(Z^2) = M_Z''(0) = 1.$$

We also compute the third moment

$$M_Z^{(3)}(t) = te^{t^2/2} + 2te^{t^2/2} + t^3 e^{t^2/2}.$$

We get

$$E(Z^3) = M_Z^{(3)}(0) = 0.$$

Example 7. We now use the computation in Example 5 to compute the mgf of a normal random variable X with mean μ and standard deviation σ. First note that the random variable Z defined as

$$Z = \frac{X - \mu}{\sigma}$$

is a standard normal distribution. We have that

$$M_X(t) = M_{\sigma Z + \mu}(t) = E(e^{t(\sigma Z + \mu)}).$$

Observe that $e^{t\mu}$ is a constant with respect to the expectation so

$$M_X(t) = e^{t\mu} E(e^{t\sigma Z}) = M_Z(t\sigma).$$

We now use that $M_Z(t) = e^{t^2/2}$ to get

$$M_X(t) = \exp(t\mu)\exp(t^2\sigma^2/2) = \exp(t\mu + t^2\sigma^2/2).$$

Our next example deals with the Gamma distribution.

Example 8. A random variable X is said to have a Gamma distribution with parameters $r > 0$ and $\lambda > 0$ if its density is

$$f(x) = e^{-\lambda x}\frac{\lambda^r x^{r-1}}{\Gamma(r)} \text{ for } x > 0$$

where

$$\Gamma(r) = \int_0^\infty x^{r-1}e^{-x}dx.$$

The improper integral above is convergent for all $r > 0$. Moreover, an easy induction proof shows that

$$\Gamma(n) = (n-1)! \text{ for all integers } n \geq 1.$$

Observe that a Gamma random variable with parameters $r = 1$ and λ is an exponential random variable with parameter λ. We now compute its mgf,

$$M_X(t) = E(e^{tX}) = \int_0^\infty e^{tx}e^{-\lambda x}\frac{\lambda^r x^{r-1}}{\Gamma(r)}dx.$$

The preceding improper integral converges only for $t < \lambda$. We divide and multiply the integrand by $(\lambda - t)^r$ to generate another Gamma distribution.

$$M_X(t) = \frac{\lambda^r}{(\lambda-t)^r}\int_0^\infty e^{-(\lambda-t)x}(\lambda-t)^r\frac{x^{r-1}}{\Gamma(r)}dx.$$

But $g(x) = e^{-(\lambda-t)x}(\lambda-t)^r\frac{x^{r-1}}{\Gamma(r)}$ is the density of a Gamma random variable with parameters r and $\lambda - t$. Thus,

$$\int_0^\infty e^{-(\lambda-t)x}(\lambda-t)^r\frac{x^{r-1}}{\Gamma(r)}dx = 1$$

and

$$M_X(t) = \frac{\lambda^r}{(\lambda-t)^r} \text{ for } t < \lambda.$$

Exercises

1. Compute the moment generating function of a geometric random variable with parameter p.

2. Compute the mgf of a uniform random variable on $[0,1]$.

3. Compute the first three moments of a binomial random variable by taking derivatives of its mgf.

4. Compute the first two moments of a geometric random variable by using Exercise 1.

5. Compute the first two moments of a uniform random variable on [0,1] by using Exercise 2.

6. Use the mgf in Example 8 to compute the mean and standard deviation of a Gamma distribution with parameters n and λ.

7. What is the mgf of a normal distribution with mean 1 and standard deviation 2?

8. Use the mgf in Example 7 to compute the first two moments of a normal distribution with mean μ and standard deviation σ.

9. (a) Make a change of variables to show that

$$\int_0^\infty e^{-\lambda x} \lambda^r x^{r-1} dx = \Gamma(r).$$

(b) Show that for all $r > 0$ and $\lambda > 0$,

$$\int_0^\infty e^{-\lambda x} \frac{\lambda^r x^{r-1}}{\Gamma(r)} dx = 1.$$

10. A random variable with density

$$f(x) = \frac{1}{2^{n/2}\Gamma(n/2)} x^{n/2-1} e^{-x/2}$$

is said to be a Chi-square random variable with n degrees of freedom (n is an integer). Find the moment generating function of X.

7.2   Sums of Independent Random Variables

We first summarize the mgf we have computed in Section 7.1.

Random Variable	Moment generating function
Binomial (n, p)	$(pe^t + 1 - p)^n$
Poisson (λ)	$\exp(\lambda(-1 + e^t))$
Exponential (λ)	$\frac{\lambda}{\lambda-t}$ for $t < \lambda$
Normal (μ, σ^2)	$\exp(t\mu + t^2\sigma^2/2)$
Gamma (r, λ)	$\frac{\lambda^r}{(\lambda-t)^r}$ for $t < \lambda$

We will use moment generating functions to show the following important property of normal random variables.

Linear Combination of Independent Normal Random Variables

Assume that $X_1, X_2, \ldots, X_n$ are independent normal random variables with mean μ_i and variance σ_i^2. Let $a_1, a_2, \ldots, a_n$ be a sequence of real numbers. Then

$$a_1 X_1 + a_2 X_2 + \cdots + a_n X_n$$

is also a normal variable with mean

$$a_1 \mu_1 + a_2 \mu_2 + \cdots + a_n \mu_n$$

and variance

$$a_1^2 \sigma_1^2 + a_2^2 \sigma_2^2 + \cdots + a_n^2 \sigma_n^2.$$

The remarkable fact here is that a linear combination of independent normal random variables is normal.

Example 1. Assume that in a given population, heights are normally distributed. The mean height for men is 172 cm with SD 5 cm and for women the mean is 165 cm with SD 3 cm. What is the probability that a woman taken at random is taller than a man taken at random?

Let X be the man's height and let Y be the woman's height. We want $P(X < Y) = P(Y - X > 0)$. According to the preceding property $Y - X$ is normally distributed with

$$E(Y - X) = E(Y) - E(X) = 165 - 172 = -7$$

and

$$Var(Y - X) = Var(Y) + Var(X) = 3^2 + 5^2 = 34.$$

We normalize $Y - X$ to get

$$P(X < Y) = P(Y - X > 0) = p\left(\frac{Y - X - (-7)}{\sqrt{34}} > \frac{0 - (-7)}{\sqrt{34}}\right)$$

$$= P\left(Z > \frac{7}{\sqrt{34}}\right) = 0.12.$$

Example 2. Assume that at a certain university, salaries of junior faculty are normally distributed with mean 40,000 and SD 5,000. Assume also that salaries of senior faculty are normally distributed with mean 60,000 and SD 10,000. What is the probability that the salary of a senior faculty member taken at random is at least twice the salary of a junior faculty member taken at random?

Let X be the salary of the junior faculty member and Y be the salary of the senior faculty member. We want $P(Y > 2X)$. We know that $Y - 2X$ is normally distributed. We express all the figures in thousands of dollars to get

$$E(Y-2X) = -20 \text{ and } Var(Y-2X) = Var(Y)+4Var(X) = 10^2+4\times5^2 = 200.$$

We normalize to get

$$P(Y - 2X > 0) = P\left(\frac{Y - 2X - (-20)}{\sqrt{200}} > \frac{0 - (-20)}{\sqrt{200}}\right)$$

$$= P\left(Z > \frac{20}{\sqrt{200}}\right) = 0.08.$$

Before proving that a linear combination of independent normally distributed random variables is normally distributed we need two properties of moment generating functions that we now state.

P1. The moment generating function of a random variable characterizes its distribution. That is, if two random variables X and Y are such that

$$M_X(t) = M_Y(t) \text{ for all } t \text{ in } (-r, r)$$

for some $r > 0$, then X and Y have the same distribution.

P1 is a crucial property. It tells us that if we recognize a moment generating function then we know what the underlying distribution is.

P2. Assume that the random variables $X_1, X_2, \ldots, X_n$ are independent and have moment generating functions. Let $S = X_1 + X_2 + \cdots + X_n$; then

$$M_S(t) = M_{X_1}(t)M_{X_2}(t) \ldots M_{X_n}(t).$$

The proof of P1 involves mathematics that are beyond the scope of this book. For a proof of P2 see P2 in Section 8.3. We now prove that a linear combination of independent normally distributed random variables is normally distributed. Assume that $X_1, X_2, \ldots, X_n$ are independent normal random variables with mean μ_i and variance σ_i^2. Let $a_1, a_2, \ldots, a_n$ be a sequence of real numbers. We compute the mgf of $a_1X_1 + a_2X_2 + \cdots + a_nX_n$. The random variables $a_i X_i$ are independent so by P2 we have

$$M_{a_1X_1+a_2X_2+\cdots+a_nX_n}(t) = M_{a_1X_1}(t)M_{a_2X_2}(t) \ldots M_{a_nX_n}(t).$$

Note that by definition

$$M_{a_iX_i}(t) = E(e^{ta_iX_i}) = M_{X_i}(a_it).$$

We now use the mgf corresponding to the normal distribution to get

$$M_{a_iX_i}(t) = \exp(a_it\mu_i + a_i^2t^2\sigma_i^2/2).$$

Thus,

$$M_{a_1 X_1 + a_2 X_2 + \cdots + a_n X_n}(t) = \exp(a_1 t \mu_1 + a_1^2 t^2 \sigma_1^2 / 2) \times \cdots \times \exp(a_n t \mu_n + a_n^2 t^2 \sigma_2^2 / 2).$$

Therefore,

$$M_{a_1 X_1 + a_2 X_2 + \cdots + a_n X_n}(t) = \exp((a_1 \mu_1 + \cdots + a_n \mu_n)t + (a_1^2 \sigma_1^2 + \cdots + a_n^2 \sigma_n^2)t^2 / 2).$$

This is exactly the mgf of a normal random variable with mean

$$a_1 \mu_1 + \cdots + a_n \mu_n$$

and variance

$$a_1^2 \sigma_1^2 + \cdots + a_n^2 \sigma_n^2.$$

So according to property P1 this shows that $a_1 X_1 + \cdots + a_n X_n$ follows a normal distribution with mean and variance given above.

Example 3. Let $T_1, \ldots, T_n$ be i.i.d. exponentially distributed random variables with rate λ. What is the distribution of $T_1 + T_2 + \cdots + T_n$?

We compute the mgf of the sum by using Property P2.

$$M_{T_1 + T_2 + \cdots + T_n}(t) = M_{T_1}(t) M_{T_2}(t) \ldots M_{T_n}(t) = M_{T_1}^n(t)$$

since all the T_i have all the same distribution and therefore the same mgf. We use the mgf for an exponential random variable with rate λ to get

$$M_{T_1 + T_2 + \cdots + T_n}(t) = (\frac{\lambda}{\lambda - t})^n.$$

This is not the mgf of an exponential distribution. However, this is the mgf of a Gamma distribution with parameters n and λ. That is, we have the following.

Sum of i.i.d. Exponential Random Variables

Let $T_1, \ldots, T_n$ be i.i.d. exponentially distributed random variables with rate λ. Then $T_1 + T_2 + \cdots + T_n$ has a Gamma distribution with parameters n and λ.

Example 4. Assume that you have two batteries that have an exponential lifetime with mean two hours. As soon as the first battery fails you replace it with a second battery. What is the probability that the batteries will last at least four hours?

The total time, T, the batteries will last is a sum of two exponential i.i.d. random variables. Therefore, T follows a Gamma distribution with parameters $n = 2$ and $\lambda = 1/2$. We use the density of a Gamma distribution (see Example 8 in 7.1 and note that $\Gamma(2) = 1$) to get

$$P(T > 4) = \int_4^\infty \lambda^2 t e^{-\lambda t} dt = 3e^{-2} = 0.41$$

where we use an integration by parts to get the second equality.

Example 5. Let X and Y be two independent Poisson random variables with means λ and μ, respectively. What is the distribution of $X + Y$?

We compute the mgf of $X + Y$. By property P2 we have that

$$M_{X+Y}(t) = M_X(t)M_Y(t).$$

Thus,

$$M_{X+Y}(t) = \exp(\lambda(-1 + e^t)) \times \exp(\mu(-1 + e^t)) = \exp((\lambda + \mu)(-1 + e^t)).$$

This is the moment generating function of a Poisson random variable with mean $\lambda + \mu$. Thus, by property P1, $X + Y$ is a Poisson random variable with mean $\lambda + \mu$.

We now state the general result.

Sum of Independent Poisson Random Variables

Let $N_1, \ldots, N_n$ be independent Poisson random variables with means $\lambda_1, \ldots, \lambda_n$, respectively. Then,

$$N_1 + N_2 + \cdots + N_n$$

is a Poisson random variable with mean

$$\lambda_1 + \lambda_2 + \cdots + \lambda_n.$$

Only a few distributions are stable under addition. Normal and Poisson distributions are two of them.

Example 6. Assume that at a given hospital there are on average two births of twins per month and one birth of triplets per year. Assume that both are Poisson random variables. What is the probability that on a given month there are four or more multiple births?

Let N_1 and N_2 be the numbers of births of twins and of triplets in a given month, respectively. Then $N = N_1 + N_2$ is a Poisson random variable with mean $\lambda = 2 + 1/12 = 25/12$. We have that

$$
\begin{aligned}
P(N \geq 4) &= 1 - P(N = 0) - P(N = 1) - P(N = 2) - P(N = 3) \\
&= 1 - e^{-\lambda} - \lambda e^{-\lambda} - \lambda^2 e^{-\lambda}/2 - \lambda^3 e^{-\lambda}/3! = 0.16.
\end{aligned}
$$

As noted before, when we sum two random variables with the same type of distribution we do not, in general, get the same distribution. Next, we will look at such an example.

Example 7. Roll two fair dice. What is the distribution of the sum?

Let X and Y be the faces shown by the two dice. The random variables X and Y are discrete uniform random variables on $\{1, 2 \ldots, 6\}$. Let $S = X + Y$. Note that S must be an integer between 2 and 12. We have that

$$P(S = 2) = P(X = 1; Y = 1) = P(X = 1)P(Y = 1) = 1/36$$

where we use the independence of X and Y to get the second equality. Likewise, we have that

$$P(S = 3) = P(X = 1, Y = 2) + P(X = 2, Y = 1) = 2/36.$$

In general, we have the following formula

$$P(S = n) = \sum_{k=1}^{n-1} P(X = k)P(Y = n - k) \text{ for } n = 2, 3 \ldots, 12.$$

We get the following distribution for S.

k	2	3	4	5	6	7	8	9	10	11	12
$P(X = k)$	1/36	2/36	3/36	4/36	5/36	6/36	5/36	4/36	3/36	2/36	1/36

Note that S is not a uniform random variable. In this case using the moment generating function does not help. We could compute the mgf of S but it would not correspond to any distribution we know.

We now state the general form of the distribution of the sum of two independent random variables.

Sum of Two Independent Random Variables

Let X and Y be two discrete independent random variables. The distribution of $X + Y$ may be computed by using the formula

$$P(X + Y = n) = \sum_{k} P(X = k)P(Y = n - k).$$

If X and Y are independent continuous random variables with densities f and g, then $X + Y$ has density h that may be computed by using the formula

$$h(x) = \int f(y)g(x - y)dy = \int g(y)f(x - y)dy.$$

We now apply the preceding formula to uniform random variables.

Example 8. Let U and V be two independent uniform random variables on $[0,1]$. The density for both of them is $f(x) = 1$ for x in $[0,1]$. Let $S = U + V$ and let h be the density of S. We have that

$$h(x) = \int f(y)f(x-y)dy.$$

Note that $f(y) > 0$ if and only if y is in $[0,1]$. Note also that $f(x-y) > 0$ if and only if $x - y$ is in $[0,1]$, that is y is in $[-1+x, x]$. Thus, $f(y)f(y-x) > 0$ if and only if y is simultaneously in $[0,1]$ and in $[-1+x, x]$. So

$$h(x) = \int_0^x dy = x \text{ if } x \text{ is in } [0, 1]$$

and

$$h(x) = \int_{-1+x}^1 dy = 2 - x \text{ if } x \text{ is in } [1, 2].$$

Observe that the sum of two uniform random variables is not uniform; the density has a triangular shape instead.

Example 9. Let X and Y be two independent exponentially distributed random variables with rates 1 and 2, respectively. What is the density of $X + Y$?

The densities of X and Y are $f(x) = e^{-x}$ for $x > 0$ and $g(x) = 2e^{-2x}$ for $x > 0$, respectively. The density h of the sum $X + Y$ is

$$h(x) = \int f(y)g(x-y).$$

In order for $f(y)g(x-y) > 0$ we need $y > 0$ and $x - y > 0$. Thus,

$$h(x) = \int_0^x e^{-y}2e^{-2(x-y)}dy \text{ for } x > 0.$$

We get

$$h(x) = 2(e^{-x} - e^{-2x}) \text{ for } x > 0.$$

Note that this is not the density of an exponential distribution. If the two rates were the same we would have obtained a Gamma distribution for the sum, but with different rates we get another type of distribution.

Proof of the Central Limit Theorem

We now sketch the proof of the Central Limit Theorem in the particular case where the random variables have moment generating functions (in the general case it is only assumed that the random variables have finite second moments). Let $X_1, X_2, \ldots, X_n$ be a sequence of independent identically distributed random variables with mean μ and variance σ^2. Assume that X_i has an mgf M_{X_i} defined on $(-r, r)$ for some $r > 0$. Let

$$\bar{X} = \frac{X_1 + X_2 + \cdots + X_n}{n}.$$

We want to show that the distribution of

$$T = \frac{\bar{X} - \mu}{\sigma/\sqrt{n}}$$

approaches the distribution of a standard normal distribution. We start by computing the moment generating function of T.

$$M_T(t) = E(e^{tT}) = E\left(\exp\left(t\frac{\bar{X} - \mu}{\sigma/\sqrt{n}}\right)\right) = E\left(\exp\left(t\sqrt{n}\frac{\bar{X} - \mu}{\sigma}\right)\right).$$

Observe now that

$$\frac{\bar{X} - \mu}{\sigma} = \frac{1}{n}\sum_{i=1}^{n}\frac{X_i - \mu}{\sigma}.$$

Let $Y_i = \frac{X_i - \mu}{\sigma}$ and $S = \sum_{i=1}^{n} Y_i$. We have that

$$M_T(t) = E\left(\exp\left(t\frac{\sqrt{n}}{n}\sum_{i=1}^{n} Y_i\right)\right) = E\left(\exp\left(t\frac{\sqrt{n}}{n}S\right)\right) = M_S\left(\frac{t}{\sqrt{n}}\right).$$

Since the Y_i are independent we get by P2 that

$$M_T(t) = M_Y\left(\frac{t}{\sqrt{n}}\right)^n.$$

We now write a third degree Taylor expansion for M_Y.

$$M_Y\left(\frac{t}{\sqrt{n}}\right) = M_Y(0) + \frac{t}{\sqrt{n}}M_Y'(0) + \frac{t^2}{2n}M_Y''(0) + \frac{t^3}{6n^{3/2}}M_Y'''(s)$$

for some s in $(0, \frac{t}{\sqrt{n}})$. Since the Y_i are standardized we have that

$$M_Y'(0) = E(Y) = 0 \text{ and } M_Y''(0) = E(Y^2) = Var(Y) = 1.$$

We also have (for any random variable) that $M_Y(0) = 1$. Thus,

$$M_Y\left(\frac{t}{\sqrt{n}}\right) = 1 + \frac{t^2}{2n} + \frac{t^3}{6n^{3/2}}M_Y'''(s).$$

We have that

$$\ln(M_T(t)) = \ln\left(M_Y\left(\frac{t}{\sqrt{n}}\right)^n\right) = n\ln\left(M_Y\left(\frac{t}{\sqrt{n}}\right)\right) = n\ln\left(1 + \frac{t^2}{2n} + \frac{t^3}{6n^{3/2}}M_Y'''(s)\right).$$

By writing that the derivative of $\ln(1 + x)$ at 0 is 1 we get

$$\lim_{x \to 0} \frac{\ln(1 + x)}{x} = 1.$$

Let $x_n = \frac{t^2}{2n} + \frac{t^3}{6n^{3/2}} M_Y'''(s)$ and note that x_n converges to 0 as n goes to infinity. We have

$$n \ln\left(M_Y\left(\frac{t}{\sqrt{n}}\right)\right) = n \ln(1 + x_n) = n x_n \frac{\ln(1 + x_n)}{x_n}.$$

Note that $M_Y'''(s)$ converges to $M_Y'''(0)$ as n goes to infinity. Thus,

$$\lim_{n \to \infty} n x_n = t^2/2 \text{ and } \lim_{n \to \infty} \frac{\ln(1 + x_n)}{x_n} = 1.$$

Therefore,

$$\lim_{n \to \infty} \ln(M_T(t)) = t^2/2 \text{ and } \lim_{n \to \infty} M_T(t) = e^{t^2/2}.$$

That is, this computation shows that the mgf of $\frac{\bar{X} - \mu}{\sigma/\sqrt{n}}$ converges to the moment generating function of a standard normal distribution. A rather deep result of probability theory called Levy's Continuity Theorem shows that this is enough to prove that the distribution of $\frac{\bar{X} - \mu}{\sigma/\sqrt{n}}$ converges to the distribution of a standard normal random variable. This concludes the sketch of the proof of the CLT.

Exercises

1. The weight of a manufactured product is normally distributed with mean 5 kg and SD 0.1 kg.

(a) Take two items at random, what is the probability that they have a weight difference of at least 0.3 kg?

(b) Take three items at random. What is the probability that the sum of the three weights is less than 14 kg?

2. Consider X a binomial random variable with parameters $n = 10$ and p. Let Y be independent of X and be a binomial random variable with $n = 15$ and p. Let $S = X + Y$.

(a) Find the mgf of S.

(b) What is the distribution of S?

3. Let X be normally distributed with mean 10 and SD 1. Let $Y = 2X - 30$.

(a) Compute the mgf of Y.

(b) Use (a) to show that Y is normally distributed and to find the mean and SD of Y.

4. Let X be the number of students from University A that get into Medical School at University B. Let Y be the number of students from University A that get into

Law School at University B. Assume that X and Y are two independent Poisson random variables with means 2 and 3, respectively. What is the probability that $X + Y$ is larger than 5?

5. Assume that 6-year old weights are normally distributed with mean 20 kg and SD 3 kg. Assume that male adults' weights are normally distributed with mean 70 kg and SD 6 kg. What is the probability that the sum of the weights of three children is larger than an adult's weight?

6. Assume you roll a die three times; you win each time you get a 6. Assume you toss a coin twice; you win each time heads comes up. Compute the distribution of your total number of wins.

7. Find the density of a sum of three independent uniform random variables on [0,1]. You may use the result for the sum of two uniform random variables in Example 8.

8. Let X and Y be two independent geometric random variables with the same probability of success p. Find the distribution of $X + Y$.

9. (a) Use moment generating functions to show that if X and Y are independent binomial random variables with parameters n and p, and m and p, respectively, then $X + Y$ is also a binomial random variable.

(b) If the probabilities of success are distinct for X and Y, is $X + Y$ a binomial random variable?

8
Transformations of Random Variables and Random Vectors

8.1 Distribution Functions and Transformations of Random Variables

Distribution functions

The notion of distribution function is especially useful when dealing with continuous random variables. However, any random variable has a distribution function, as one can see below.

Distribution Function

Let X be a random variable. The function

$$F(x) = P(X \leq x)$$

is called the distribution function of X. If X is a continuous random variable with density f, then
$$F'(x) = f(x)$$
for all the points x where f is continuous.

Recall that if X is a continuous random variable with density f, then for any a and b such that $-\infty \leq a \leq b \leq +\infty$,

$$P(a \leq X \leq b) = \int_a^b f(x)dx.$$

Therefore, if X has density f and distribution function F, then

$$F(x) = P(X \le x) = \int_{-\infty}^{x} f(t)dt.$$

By the Fundamental Theorem of Calculus, if f is continuous at x, then F is differentiable at x and

$$F'(x) = f(x).$$

The preceding equality shows that if we know the distribution function, then we know the density and therefore the distribution of a continuous random variable. This is true for any random variable: a distribution function determines the distribution of a random variable.

Next we compute a few distribution functions.

Example 1. Let U be a uniform random variable on $[0,1]$. That is, the density of U is $f(u) = 1$ for u in $[0,1]$ and $f(u) = 0$ elsewhere. The distribution function F of U is

$$F(u) = \int_{-\infty}^{u} f(x)dx.$$

In order to compute F explicitly we need to consider three cases. If $u \le 0$, then f is 0 on $(-\infty, u)$ and $F(u) = 0$. If $0 < u < 1$, then

$$F(u) = \int_{-\infty}^{u} f(x)dx = \int_{0}^{u} f(x)dx = \int_{0}^{u} dx = u.$$

Finally, if $u \ge 1$, then

$$F(u) = \int_{-\infty}^{u} f(x)dx = \int_{-\infty}^{+\infty} f(x)dx = 1.$$

This is so because if $u \ge 1$, then f is 0 on $(u, +\infty)$ and the integral of a density function on the whole line is always 1. Summarizing the computations above we get

$$
\begin{aligned}
F(u) &= 0 \text{ if } u \le 0, \\
F(u) &= u \text{ if } 0 < u < 1, \\
F(u) &= 1 \text{ if } u \ge 1.
\end{aligned}
$$

Below we sketch the graph of F.

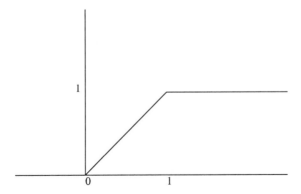

There are three features of the graph above that are typical of all distribution functions and that we now state.

Properties of Distribution Functions

Let F be the distribution function of a random variable X. Then, we have the following three properties:

$$\lim_{x \to -\infty} F(x) = 0.$$

F is an increasing function. That is, if $x_1 < x_2$ then $F(x_1) \leq F(x_2)$.

$$\lim_{x \to +\infty} F(x) = 1.$$

Example 2. Let T be an exponential random variable with rate λ. What is its distribution function?

The density of T is $f(t) = \lambda e^{-\lambda t}$ for $t \geq 0$. Thus, $F(t) = 0$ for $t \leq 0$ and for $t > 0$,

$$F(t) = \int_{-\infty}^{t} f(x)dx = \int_{0}^{t} f(x)dx = -e^{-\lambda x}]_0^t = 1 - e^{-\lambda t}.$$

It follows that $F(t) = 0$ if $t \leq 0$, $F(t) = 1 - e^{-\lambda t}$ if $t > 0$. One can easily check that the three properties of distribution functions hold here as well.

In the next example we give a first application of the notion of distribution function.

Example 3. Assume that T_1 and T_2 are two independent exponentially distributed random variables with rates λ_1 and λ_2, respectively. Let T be the minimum of T_1 and T_2. What is the distribution of T?

Let F be the distribution function of T. We have that

$$F(t) = P(T \leq t) = P(\min(T_1, T_2) \leq t) = 1 - P(\min(T_1, T_2) > t).$$

Observe that $\min(T_1, T_2) > t$ if and only if $T_1 > t$ and $T_2 > t$. Thus, since we are assuming that T_1 and T_2 are independent we get

$$F(t) = 1 - P(T_1 > t)P(T_2 > t) = 1 - (1 - F_1(t))(1 - F_2(t))$$

where F_1 and F_2 are the distribution functions of T_1 and T_2, respectively. Using the form of the distribution function given in Example 2 we get

$$F(t) = 1 - e^{-\lambda_1 t} e^{-\lambda_2 t} = 1 - e^{-(\lambda_1 + \lambda_2)t} \text{ for } t \geq 0.$$

Therefore, the computation above shows that the minimum of two independent exponential random variables is also exponentially distributed and its rate is the sum of the two rates.

Next we look at the maximum of two random variables.

Example 4. Let U_1, U_2 and U_3 be three independent random variables uniformly distributed on [0,1]. Let M be the maximum of U_1, U_2, U_3. What is the density of M?

We are going to compute the distribution function of M and then differentiate to get the density. Let F be the distribution function of M. We have that

$$F(x) = P(M \leq x) = P(\max(U_1, U_2, U_3) \leq x).$$

We have that $\max(U_1, U_2, U_3) \leq x$ if and only if $U_i \leq x$ for $i = 1, 2, 3$. Thus, due to the independence of the U_i we get

$$F(x) = P(U_1 \leq x)P(U_2 \leq x)P(U_3 \leq x) = F_1(x)F_2(x)F_3(x).$$

According to Example 1,

$$\begin{aligned}
F(x) &= 0 \text{ if } x \leq 0, \\
F(x) &= x^3 \text{ if } 0 < x < 1, \\
F(x) &= 1 \text{ if } x \geq 1.
\end{aligned}$$

Thus, the density of M, which we denote by f, is

$$f(x) = F'(x) = 3x^2 \text{ for } x \text{ in } [0, 1].$$

Observe that the maximum of uniform random variables is not a uniform random variable.

In Examples 3 and 4, in order to compute the distribution of a minimum and a maximum we have used distribution functions. This is a general method that we may summarize below.

Maximum and Minimum of Independent Random Variables

Let $X_1, X_2, \ldots, X_n$ be independent random variables with distribution functions $F_1, F_2, \ldots, F_n$, respectively. Let F_{max} and F_{min} be the distribution functions of the random variables $\max(X_1, X_2, \ldots, X_n)$ and $\min(X_1, X_2, \ldots, X_n)$, respectively. Then,

$$F_{max} = F_1 F_2 \ldots F_n$$

and

$$F_{min} = 1 - (1 - F_1)(1 - F_2) \ldots (1 - F_n).$$

We now give an example of a distribution function for a discrete random variable.

Example 5. Flip two fair coins and let X be the number of tails. The distribution of X is given by

x	0	1	2
$P(X = x)$	1/4	1/2	1/4

Observe that if $0 \le x < 1$, then

$$F(x) = P(X \le x) = P(X = 0) = 1/4,$$

while if $1 \le x < 2$, then

$$F(x) = P(X \le x) = P(X = 0) + P(X = 1) = 3/4.$$

Therefore the distribution function of this discrete random variable is given by

$$
\begin{aligned}
F(x) &= 0 \text{ if } x < 0,\\
F(x) &= 1/4 \text{ if } 0 \le x < 1,\\
F(x) &= 3/4 \text{ if } 1 \le x < 2,\\
F(x) &= 1 \text{ if } x \ge 2.
\end{aligned}
$$

The graph is given below.

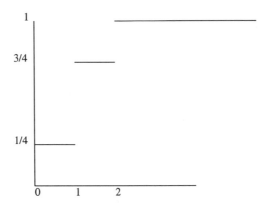

Note that distribution functions of discrete random variables are always discontinuous.

Simulations

Simulations are another type of application of distribution functions. We will see how one can simulate a random variable with a given distribution by using a simulation of a uniform random variable. Many computer random simulators create numbers that behave approximately as observations of independent uniform random variables on [0,1]. So our problem is to go from a uniform random variable to another distribution. We start by considering a continuous random variable X. Assume that the distribution function F of X is strictly increasing and continuous so that the inverse function F^{-1} is well defined. Let U be a uniform random variable on [0,1]. We have that

$$P(F^{-1}(U) \leq x) = P(U \leq F(x)) = F(x)$$

since $F(x)$ is always in [0,1] and $P(U \leq x) = x$ for x in [0,1]. This shows the following.

Simulation of a Continuous Random Variable

Let X be a continuous random variable with a strictly increasing distribution function F. Let U be a uniform random variable on [0,1]. Then $F^{-1}(U)$ has the same distribution as X. That is, to simulate X it is enough to simulate a uniform random variable U and then compute $F^{-1}(U)$.

Example 6. A computer random simulator gives us the following ten random numbers: 0.38,0.1,0.6,0.89,0.96,0.89,0.01,0.41,0.86,0.13. Simulate ten independent exponential random variables with rate 1.

By Example 2 we know that the distribution function F of an exponential random variable with rate 1 is

$$F(x) = 1 - e^{-x}.$$

We compute F^{-1}. If

$$y = 1 - e^{-x},$$

then

$$x = - \ln(1 - y).$$

Thus,

$$F^{-1}(x) = - \ln(1 - x).$$

We now compute $F^{-1}(x)$ for $x = 0.38, 0.1, 0.6, 0.89, 0.96, 0.89, 0.01, 0.41, 0.86, 0.13$. We get the following ten observations for ten independent exponential rate 1 random variables: 4.78, 1.05, 0.92, 2.21, 3.22, 2.21, 4.6, 5.28, 1.97, 1.39.

We now turn to the simulation of discrete random variables. Consider a discrete random variable X with k values: 0,1,2, ... ,k. Denote $P(X = i) = p_i$ for $i = 1, 2, \ldots, k$. Let U be a uniform random variable. The following algorithm uses a simulation of U to give a simulation of X.

$$\text{If } U < p_0 \text{ set } X = 0.$$

$$\text{If } p_0 \leq U < p_0 + p_1 \text{ set } X = 1.$$

More generally, for $i = 1, 2 \ldots, k$,

$$\text{if } \quad p_0 + p_1 + \cdots + p_{i-1} \leq U < p_0 + p_1 + \cdots + p_{i-1} + p_i, \quad \text{set } X = i.$$

Recall that for $0 \leq a \leq b \leq 1$ we have

$$P(a \leq U \leq b) = b - a.$$

Thus,

$$P(X = 0) = P(U < p_0) = p_0 \text{ and } P(X = 1) = P(p_0 \leq U < p_0 + p_1) = p_1.$$

In general this algorithm yields $P(X = i) = p_i$ for $i = 1, 2, \ldots, k$. That is, we are able to simulate X from a simulation of U. We now use this algorithm in an example.

Example 7. Let X be a binomial with parameters $n = 2$ and $p = 1/2$. The distribution of X is given by $p_0 = 1/4$, $p_1 = 1/2$ and $p_2 = 1/4$. A random generator gives us the following random numbers: 0.38,0.1,0.6,0.89,0.96,0.89,0.01,0.41,0.86,0.13. Note that

$$p_0 \leq 0.38 \leq p_0 + p_1.$$

Thus, the first simulation for the random variable X is $X = 1$. Using the algorithm above we get the following simulation of 10 independent random variables with the same distribution as X: 1,0,1,2,2,2,0,1,2,0.

Transformations of random variables

At this point we know relatively few different continuous distributions: uniform, exponential and normal are the main distributions we have seen. In this section we will see a general method to obtain many more distributions from the known ones. We start with an example.

Example 8. Let U be a uniform random variable on $[0,1]$. Define $X = U^2$. What is the distribution of X?

We use the distribution function F of X.

$$F(x) = P(X \leq x) = P(U^2 \leq x) = P(U \leq \sqrt{x}).$$

Recall that $P(U \leq y) = y$ for y in $[0,1]$. Thus,

$$F(x) = \sqrt{x} \text{ for } 0 \leq x \leq 1.$$

From this we may compute the density of X:

$$f(x) = F'(x) = \frac{1}{2\sqrt{x}} \text{ for } 0 \leq x \leq 1.$$

More generally, we may use distributions functions to compute distributions of functions of random variables. Assume that a continuous random variable X has values on (a, b) and has density f_X. Define a new random variable $Y = h(X)$ where h is a real valued strictly monotone function defined on (a, b). Let F_Y be the distribution function of Y. We have

$$F_Y(y) = P(Y \leq y) = P(h(X) \leq y) = P(X \leq h^{-1}(y)) = F_X(h^{-1}(y)).$$

Thus, we have the following simple relation between the distribution functions of X and Y:

$$F_Y = F_X \circ h^{-1}.$$

Assuming that h is differentiable we take derivatives on both sides to get a relation between the densities of X and Y:

$$f_Y = (f_X \circ h^{-1}) \times \frac{1}{h' \circ h^{-1}}.$$

Example 9. We redo Example 1 by using the formula above. We have that $h(x) = x^2$ is strictly increasing on $(0,1)$ and $h^{-1}(x) = \sqrt{x}$. Note that $f_X(x) = 1$ for x in $(0,1)$, thus $f_X(\sqrt{y}) = 1$ for y in $(0,1)$. We have that $h'(x) = 2x$. Therefore,

$$f_Y(y) = \frac{f_X(h^{-1}(y))}{h'(h^{-1}(y))} = \frac{1}{2\sqrt{y}} \text{ for } 0 \leq y \leq 1.$$

We now summarize the method.

Transformation of a Random Variable

Assume that a continuous random variable X has values on (a, b) and has density f_X. Define a new random variable $Y = h(X)$ where h is a real valued differentiable strictly monotone function defined on (a, b). We have the following relation between the distribution functions F_X and F_Y of X and Y.

$$F_Y(y) = P(Y \le y) = P(h(X) \le y) = P(X \le h^{-1}(y)) = F_X(h^{-1}(y)).$$

We also have a relation between the two density functions f_X and f_Y:

$$f_Y(y) = \frac{f_X(h^{-1}(y))}{h'(h^{-1}(y))}.$$

As the following example shows one has to check carefully that the hypotheses above hold before using the formula. In fact, rather than memorizing the above formulas the reader should do the computation every time as was done in Example 1.

Example 10: The Chi-Square distribution. Let Z be a standard normal random variable. What is the density of $Y = Z^2$?

We have for $y \ge 0$,

$$F_Y(y) = P(Y \le y) = P(Z^2 \le y) = P(-\sqrt{y} \le Z \le \sqrt{y}) = F_Z(\sqrt{y}) - F_Z(-\sqrt{y}).$$

Recall that the density of Z is

$$f_Z(z) = \frac{1}{\sqrt{2\pi}} e^{-z^2/2}.$$

Using the chain rule we get

$$f_Y(y) = f_Z(\sqrt{y}) \times \frac{1}{2\sqrt{y}} - f_Z(-\sqrt{y}) \times \frac{-1}{2\sqrt{y}}.$$

Hence, the density of Y is

$$f_Y(y) = \frac{1}{\sqrt{2\pi}} y^{-1/2} e^{-y/2} \text{ for } y \ge 0.$$

This is the density of the Chi-Square distribution with one degree of freedom. Note that the transformation $h(x) = x^2$, from the real numbers into the real numbers, is *not* one-to-one. One cannot use the formulas above directly. In this case it is best to perform a direct computation as the one we just did.

Example 11. Let T be exponentially distributed with mean 1. What is the distribution of $X = \sqrt{T}$?

Take $x \geq 0$.

$$F_X(x) = P(X \leq x) = P(\sqrt{T} \leq x) = P(T \leq x^2) = \int_0^{x^2} e^{-t} dt = 1 - e^{-x^2}.$$

Thus, the density of X is

$$f_X(x) = 2xe^{-x^2} \text{ for } x \geq 0.$$

Next we prove a property of normal random variables that we have already used many times.

Example 12. Let X be a normal random variable with mean μ and standard deviation σ. Show that $Y = \frac{X-\mu}{\sigma}$ is a standard normal random variable.

We compute the distribution function of Y:

$$F_Y(y) = P(Y \leq y) = P\left(\frac{X-\mu}{\sigma} \leq y\right) = P(X \leq \mu + \sigma y) = F_X(\mu + \sigma y).$$

We take derivatives to get

$$f_Y(y) = f_X(\mu + \sigma y) \times \sigma.$$

Recall that the density of X is

$$f_X(x) = \frac{1}{\sigma\sqrt{2\pi}} e^{-\frac{(x-\mu)^2}{2\sigma^2}}.$$

Thus,

$$f_Y(y) = \frac{1}{\sqrt{2\pi}} e^{-y^2/2}.$$

This proves that Y is a standard normal distribution.

Exercises

1. Compute the distribution function of a uniform random variable on $[-1, 2]$.

2. Assume that waiting times for buses from lines 5 and 8 are exponentially distributed with means 10 and 20 minutes, respectively. I can take either line so I will take the first bus that comes.

(a) Compute the probability that I will have to wait at least 15 minutes?

(b) What is the mean time I will have to wait?

3. Consider a circuit with two components in parallel. Assume that both components have independent exponential lifetimes with means 1 and 2 years, respectively.
(a) What is the probability that the circuit lasts more than three years?
(b) What is the expected lifetime of the circuit?

4. Assume that T_1 and T_2 are two independent exponentially distributed random variables with rates λ_1 and λ_2, respectively. Let M be the maximum of T_1 and T_2; what is the density of M?

5. Roll a fair die. Let X be the face shown. Graph the distribution function of X.

6. Consider a standard normal random variable Z. Use a normal table to sketch the graph of the distribution function of Z.

7. Use the random numbers from Example 6 to simulate a standard normal distribution.

8. Use the random numbers from Example 6 to simulate a Poisson distribution with mean 1.

9. Let X be a random variable with distribution function $F(x) = x^2$ for x in $[0,1]$.
(a) What is $P(X < 1/3) =?$
(b) What is the expected value of X?

10. Let $U_1, U_2, \ldots, U_n$ be n i.i.d. uniform random variables on $[0,1]$.
(a) Find the density of the maximum of the U_i.
(b) Find the density of the minimum of the U_i.

11. Let U be a uniform random variable on $[0,1]$. Define $X = \sqrt{U}$. What is the density of X?

12. Let T be exponentially distributed with mean 1. What is the expected value of $T^{1/3}$?

13. Let Z be a standard normal distribution. Find the density of $X = e^Z$. (X is called a lognormal random variable).

14. Let U be uniform on $[0,1]$. Find the density of $Y = \ln(1 - U)$.

15. Let T be exponentially distributed with rate λ. Find the density of $T^{1/a}$ where $a > 0$. ($T^{1/a}$ is called a Weibull random variable with parameters a and λ).

16. Let X be a continuous random variable, let $a > 0$ and b be two real numbers and let $Y = aX + b$.
(a) Show that
$$f_Y(y) = \frac{1}{a} f_X\left(\frac{y - b}{a}\right).$$
(b) Show that if X is normally distributed, then so is $Y = aX + b$.
(c) If X is exponentially distributed, is $Y = aX + b$ also exponentially distributed?

17. Consider the discrete random variable X with the following distribution.

x	-2	-1	2
$P(X = x)$	$1/4$	$1/2$	$1/4$

Find the distribution $Y = X^2$.

8.2   Random Vectors

In this section we introduce the notion of random vectors and joint distributions.

Density of a Continuous Random Vector

Let X and Y be two continuous random variables. The density of the vector (X, Y) is a positive function f such that

$$\int_{-\infty}^{+\infty} \int_{-\infty}^{+\infty} f(x, y)dxdy = 1.$$

For $a < b$ and $c < d$,

$$P(a < X < b; c < Y < d) = \int_{a}^{b} \int_{c}^{d} f(x, y)dxdy.$$

More generally, for a function g,

$$E(g(X, Y)) = \int_{-\infty}^{+\infty} \int_{-\infty}^{+\infty} g(x, y)f(x, y)dxdy,$$

provided the expectation of $g(X, Y)$ exists.

Example 1. Assume that (X, Y) is uniformly distributed on the disc $C = \{(x, y) : x^2 + y^2 \leq 1\}$. What is the density of (X, Y)?

Since we want a uniform distribution, we let $f(x, y) = c$ for (x, y) in C, $f(x, y) = 0$ elsewhere. We want

$$\int \int_{C} f(x, y) = 1 = c \times \text{area}(C).$$

Thus, $c = 1/\pi$.

At this point the reader may want to review Fubini's Theorem from calculus. It gives sufficient conditions to integrate multiple integrals one variable at a time.

Note that if the random vector (X, Y) has a density f, then for any $a < b$ we have

$$P(a < X < b) = P(a < X < b; -\infty < Y < +\infty) = \int_a^b \int_{-\infty}^{+\infty} f(x, y)dxdy.$$

Let

$$f_X(x) = \int_{-\infty}^{+\infty} f(x, y)dy;$$

then

$$P(a < X < b) = \int_a^b f_X(x)dx.$$

That is, f_X is the density of X. We now state this result.

Marginal Densities

Let (X, Y) be a random vector with density f. Then the densities of X and Y are denoted respectively by f_X and f_Y and are called the marginal densities. They are given by

$$f_X(x) = \int_{-\infty}^{+\infty} f(x, y)dy \text{ and } f_Y(y) = \int_{-\infty}^{+\infty} f(x, y)dx.$$

Example 2. We consider again the uniform random vector on the unit disc from Example 1. What are the marginals of X and Y?

Since $x^2 + y^2 \leq 1$, if we fix x in $[-1, 1]$, then y varies between $-\sqrt{1 - x^2}$ and $+\sqrt{1 - x^2}$. Thus,

$$f_X(x) = \int_{-\infty}^{+\infty} f(x, y)dy = f_X(x) = \int_{-\sqrt{1-x^2}}^{+\sqrt{1-x^2}} 1/\pi dy.$$

Therefore,

$$f_X(x) = \frac{2}{\pi}\sqrt{1 - x^2} \text{ for } x \text{ in } [-1, 1].$$

By symmetry we get that

$$f_Y(y) = \frac{2}{\pi}\sqrt{1 - y^2} \text{ for } y \text{ in } [-1, 1].$$

Note that although the vector (X, Y) is uniform, X and Y are not uniform random variables.

Recall that two random variables X and Y are said to be independent if for any $a < b$ and $c < d$ we have

$$P(a < X < b; c < Y < d) = P(a < X < b)P(c < Y < d).$$

This definition translates nicely into a property of densities that we now state.

Independence

Let (X, Y) be a random vector with density f and marginal densities f_X and f_Y. The random variables X and Y are independent if and only if

$$f(x, y) = f_X(x) f_Y(y).$$

Example 3. We continue to analyze the uniform distribution on a disc from Example 1. Are X and Y independent?

Recall that in this case we have $f(x, y) = 1/\pi$ on $C = \{(x, y) : x^2 + y^2 \leq 1\}$ and 0 elsewhere. We computed f_X and f_Y in Example 2 and clearly $f(x, y) \neq f_X(x) f_Y(y)$. We conclude that X and Y are not independent.

Example 4. Consider two electronic components that have independent exponential lifetimes with means 1 and 2 years, respectively. What is the probability that component 1 outlasts component 2?

Let T and S be respectively the lifetimes of components 1 and 2. We want $P(T > S)$. In order to compute this type of probability we need the joint distribution of (T, S). Since the two random variables are assumed to be independent we have that the joint density is

$$f(t, s) = f_T(t) f_S(s) = e^{-t} e^{-s/2}/2 \text{ for } t \geq 0, s \geq 0.$$

We now compute

$$P(T > S) = \int_{s=0}^{\infty} \int_{t=s}^{\infty} e^{-t} e^{-s/2}/2 \, dt \, ds.$$

We first integrate in t and then in s to get

$$P(T > S) = \int_{s=0}^{\infty} e^{-s} e^{-s/2}/2 \, ds = 1/3.$$

Example 5. Assume that my arrival time at the bus stop is uniformly distributed between 7:00 and 7:05. Assume that the arrival time of the bus I want to take is uniformly distributed between 7:02 and 7:04. What is the probability that I catch the bus?

To simplify the notation we do a translation of seven hours. Let U be my arrival time; it is uniformly distributed on $[0,5]$. Let V be the arrival time of the bus; it is uniformly distributed on $[2,4]$. We want the probability $P(U < V)$. It is natural to assume that U and V are independent. So we get

$$P(U < V) = \int_{v=2}^{4} \int_{u=0}^{v} \frac{1}{2} \times \frac{1}{5} \, du \, dv = \int_{v=2}^{4} \frac{v}{10} \, dv = 3/5.$$

We now turn to a discrete example.

Example 6. Let X and Y be two random variables with the following joint distribution.

X Y	0	1	2
1	1/8	1/8	1/4
2	1/8	0	1/8
3	1/8	1/8	0

By replacing integrals by sums we get the marginals of X and Y in a way that is analogous to the continuous case. To get the distribution of X we sum the joint probabilities from top to bottom.

X	0	1	2
P(X = x)	3/8	1/4	3/8

To get the distribution of Y we sum the joint probabilities from left to right.

Y	1	2	3
P(Y = y)	1/2	1/4	1/4

Two discrete random variables X and Y are independent if and only if

$$P(X = x, Y = y) = P(X = x)P(Y = y) \text{ for all } x, y.$$

In this example we see that X and Y are not independent since

$$P(X = 1, Y = 2) = 0 \text{ and } P(X = 1)P(Y = 2) = 3/16.$$

Proof that the expectation is linear

We start by proving the addition formula for expectation that we have already used many times. Assume that X and Y are continuous random variables with joint density f and that their expectations exist. Then, by using the linearity of the integral we get

$$
\begin{aligned}
E(X + Y) &= \int_{x=-\infty}^{\infty} \int_{y=-\infty}^{\infty} (x + y) f(x, y) dx dy \\
&= \int_{x=-\infty}^{\infty} \int_{y=-\infty}^{\infty} x f(x, y) dx dy + \int_{x=-\infty}^{\infty} \int_{y=-\infty}^{\infty} y f(x, y) dx dy.
\end{aligned}
$$

Note that

$$
\begin{aligned}
\int_{x=-\infty}^{\infty} \int_{y=-\infty}^{\infty} xf(x, y)dxdy &= \int_{x=-\infty}^{\infty} x\left(\int_{y=-\infty}^{\infty} f(x, y)dy\right) \\
&= \int_{x=-\infty}^{\infty} xf_X(x)dx = E(X).
\end{aligned}
$$

Similarly, we have that

$$
\int_{x=-\infty}^{\infty} \int_{y=-\infty}^{\infty} yf(x, y)dxdy = E(Y).
$$

Hence,

$$
E(X + Y) = E(X) + E(Y).
$$

For any constant a and random variable X we have

$$
E(aX) = \int axf_X(x)dx = a \int xf_X(x)dx = aE(X).
$$

Therefore, we have proved the following (the proof is analogous for discrete random variables):

The Expectation is a Linear Operator

Let X and Y be random variables; then

$$
E(X + Y) = E(X) + E(Y).
$$

For any constant a,

$$
E(aX) = aE(X).
$$

Covariance

As we will see, covariance and correlation are measures of the joint variations of X and Y.

Covariance and Correlation

Assume that X and Y are two random variables such that $E(X^2)$ and $E(Y^2)$ exist. The covariance of X and Y is

$$Cov(X, Y) = E[(X - E(X))(Y - E(Y))].$$

A computational formula for the covariance is

$$Cov(X, Y) = E(XY) - E(X)E(Y).$$

The correlation of X and Y is

$$Corr(X, Y) = \frac{Cov(X, Y)}{SD(X)SD(Y)}.$$

Correlations are standardized covariances: correlations are always in $[-1, 1]$. So it is easier to interpret a correlation than a covariance.

We prove the computational formula above.

$$Cov(X, Y) = E[(X-E(X))(Y-E(Y))] = E[XY-XE(Y)-E(X)Y+E(X)E(Y)].$$

Recalling that $E(X)$ and $E(Y)$ are constants and that the expectation is linear we get

$$\begin{aligned}
Cov(X, Y) &= E[XY] - E[XE(Y)] - E[E(X)Y] + E[E(X)E(Y)] \\
&= E(XY) - E(X)E(Y) - E(X)E(Y) + E(X)E(Y) \\
&= E(XY) - E(X)E(Y).
\end{aligned}$$

Example 7. Let (X, Y) be uniformly distributed on the triangle $\mathcal{T} = \{(x, y) : 0 < y < x < 2\}$. It is easy to see that the density of (X, Y) is $f(x, y) = 1/2$ for (x, y) in $\mathcal{T}$ and 0 elsewhere. We start by computing the marginal densities of X and Y.

$$f_X(x) = \int_0^x 1/2 \, dy = \frac{1}{2}x \text{ for } x \text{ in } [0, 2].$$

$$f_Y(y) = \int_y^2 1/2 \, dx = \frac{1}{2}(2 - y) \text{ for } y \text{ in } [0, 2].$$

We note that X and Y are not uniformly distributed and are not independent. We now compute the expectations and standard deviations of X and Y.

$$E(X) = \int_0^2 x\frac{1}{2}x \, dx = 4/3.$$

We have that

$$E(X^2) = \int_0^2 x^2 \frac{1}{2} x \, dx = 2.$$

Thus,

$$Var(X) = E(X^2) - E(X)^2 = 2/9.$$

We have

$$E(Y) = \int_0^2 y \frac{1}{2} (2 - y) \, dy = 2/3$$

and

$$E(Y^2) = \int_0^2 y^2 \frac{1}{2} (2 - y) \, dy = 2/3.$$

Thus,

$$Var(Y) = 2/9.$$

We still need to compute

$$E(XY) = \int_{x=0}^2 \int_{y=0}^x xyf(x, y) \, dx \, dy = \frac{1}{2} \int_{x=0}^2 x(x^2/2) \, dx = 1.$$

We now may compute the covariance and correlation of X and Y.

$$Cov(X, Y) = E(XY) - E(X)E(Y) = 1 - (4/3) \times (2/3) = 1/9.$$

The correlation between X and Y is

$$Corr(X, Y) = \frac{Cov(X, Y)}{SD(X)SD(Y)} = \frac{1/9}{\sqrt{2/9}\sqrt{2/9}} = 1/2.$$

A positive correlation indicates that when one variable is large it is likely that the other variable will be large as well. Next we give a few properties of correlations.

Properties of Correlations

For any random variables X and Y the correlation between X and Y is always in $[-1, 1]$. The correlation between X and Y is -1 or 1 if and only if there are constants a and b such that $Y = aX + b$. A positive correlation indicates that when one random variable is large the other one tends to be large too. Conversely, a negative correlation indicates that when one random variable is large the other one tends to be small. If $Corr(X, Y) = 0$, then X and Y are said be uncorrelated. If X and Y are independent, then they are uncorrelated. However, uncorrelated random variables do not need to be independent.

We now show that independent random variables are uncorrelated. Let X and Y be independent. Thus,

$$f(x, y) = f_X(x) f_Y(y).$$

By using the independence property above we get

$$\begin{aligned} E(XY) &= \int_{x=-\infty}^{\infty} \int_{y=-\infty}^{\infty} xy f(x, y) dx dy \\ &= \int_{x=-\infty}^{\infty} f_X(x) dx \int_{y=-\infty}^{\infty} f_Y(y) dy = E(X)E(Y). \end{aligned}$$

Therefore $Cov(X, Y) = 0$ and $Corr(X, Y) = 0$. As the next example shows, uncorrelated random variables do not need to be independent.

Example 8. We go back to the uniform random vector on the disc $C = \{(x, y) : x^2 + y^2 \le 1\}$. We have shown already in Example 3 that X and Y are not independent. However, we will show now that they are uncorrelated.

$$E(XY) = \int_{x=-1}^{1} \int_{y=-\sqrt{1-x^2}}^{\sqrt{1-x^2}} xy \frac{1}{\pi} dy dx.$$

Note that when we integrate in y we get

$$\int_{y=-\sqrt{1-x^2}}^{\sqrt{1-x^2}} y dy = 0.$$

Therefore $E(XY) = 0$. On the other hand from Example 2 we have that the density for X,

$$f_X(x) = \frac{2}{\pi} \sqrt{1 - x^2} \text{ for } x \text{ in } [-1, 1].$$

We compute

$$E(X) = \int_{-1}^{1} x f_X(x) dx = \frac{2}{\pi} (-1/2)(1 - x^2)^{3/2}]_{x=-1}^{1} = 0.$$

By symmetry we also have that $E(Y) = 0$. Therefore,

$$Cov(X, Y) = Corr(X, Y) = 0,$$

although X and Y are not independent. This is so because correlation measures the strength of the *linear* relation between X and Y. Here there is no linear relation but the random variables are related in some other way.

We now use the notion of covariance to compute the variance of a sum of random variables. Recall that

$$Var(X) = E[(X - E(X))^2].$$

Therefore, for any two random variables X and Y we have

$$\begin{aligned} Var(X + Y) &= E[(X + Y - E(X + Y))^2] \\ &= E[(X - E(X))^2 + 2(X - E(X))(Y - E(Y)) + (Y - E(Y))^2]. \end{aligned}$$

By using the linearity of the expectation and the definition of the covariance we get

$$Var(X + Y) = Var(X) + Var(Y) + 2Cov(X, Y).$$

Variance of a Sum

For any random variables X and Y,

$$Var(X + Y) = Var(X) + Var(Y) + 2Cov(X, Y)$$

provided $Var(X)$ and $Var(Y)$ exist. In particular

$$Var(X + Y) = Var(X) + Var(Y)$$

if and only if X and Y are uncorrelated.

Transformations of random vectors

A consequence of multivariate calculus is the following formula for the density of a transformed random vector.

Density of a Transformed Random Vector

Let (X, Y) be a random vector with density f. Let (U, V) be such that

$$U = g_1(X, Y) \text{ and } V = g_2(X, Y).$$

Assume that the transformation $(x, y) \longrightarrow (g_1(x, y), g_2(x, y))$ is one-to-one with inverse

$$X = h_1(U, V) \text{ and } Y = h_2(U, V).$$

Then the density of the transformed random vector (U, V) is

$$f(h_1(u, v), h_2(u, v))|J(u, v)|$$

where $J(u, v)$ is the determinant

$$\begin{vmatrix} \partial h_1/\partial u & \partial h_1/\partial v \\ \partial h_2/\partial u & \partial h_2/\partial v \end{vmatrix}.$$

We now use the preceding formula on an example.

Example 9. Let X and Y be two independent standard normal distributions. Let $U = X/Y$ and $V = X$. What is the density of (U, V)?

We see that $(x, y) \longrightarrow (u, v)$ is a one-to-one transformation from $\mathbf{R}^2$ on to $\mathbf{R}^2$. We invert the transformation to get

$$X = V \text{ and } Y = V/U.$$

We now compute the Jacobian,

$$J(u, v) = \begin{vmatrix} 0 & 1 \\ -v/u^2 & 1/u \end{vmatrix} = v/u^2.$$

Since we assume that X and Y are independent standard normal distributions we have

$$f(x, y) = \frac{1}{\sqrt{2\pi}} e^{-x^2/2} \frac{1}{\sqrt{2\pi}} e^{-y^2/2}.$$

Therefore, the density of (U, V) is

$$\frac{1}{2\pi} e^{-v^2/2} e^{-v^2/(2u^2)} |J(u, v)| = \frac{1}{2\pi} \exp\left(\frac{-v^2}{2}(1 + 1/u^2)\right) |v|/u^2.$$

We may now use this joint density to get the marginal density of U. We integrate the density above in v to get

$$f_U(u) = \int_{-\infty}^{\infty} \frac{1}{2\pi} \exp\left(\frac{-v^2}{2}(1 + 1/u^2)\right) |v|/u^2 dv.$$

Observe that the integrand above is an even function of v. Thus,

$$f_U(u) = 2 \int_0^{\infty} \frac{1}{2\pi} \exp\left(\frac{-v^2}{2}(1 + 1/u^2)\right) v/u^2 dv$$

$$= -\frac{1}{\pi} \frac{1}{1+1/u^2} \frac{1}{u^2} \exp\left(\frac{-v^2}{2}(1 + 1/u^2)\right) \Big]_{v=0}^{\infty}.$$

Hence,

$$f_U(u) = \frac{1}{\pi} \frac{1}{1 + u^2}.$$

Therefore, the ratio of two standard normal random variables follows the density above which is called the Cauchy density. Note that $E(U)$ does not exist (see Exercise 10).

Example 10. Let X and Y be two exponential and independent random variables with rates a and b respectively. Let $U = \min(X, Y)$ and $V = \max(X, Y)$. What is the joint density of (U, V)?

Note that if $X < Y$, then $U = X$ and $V = Y$. The Jacobian is then 1. The portion of the density of (U, V) corresponding to the domain $X < Y$ is then

$$ae^{-au}be^{-bv} \text{ for } 0 < u < v.$$

If $X > Y$, then $U = Y$ and $V = X$. Again the Jacobian is 1. The portion of the density of (U, V) corresponding to the domain $X > Y$ is

$$ae^{-av}be^{-bu} \text{ for } 0 < u < v.$$

We add the two parts to get the joint density of (U, V):

$$ae^{-au}be^{-bv} + ae^{-av}be^{-bu} \text{ for } 0 < u < v.$$

Are U and V independent?
 We compute

$$
\begin{aligned}
f_U(u) &= \int_{v=u}^{\infty} (ae^{-au}be^{-bv} + ae^{-av}be^{-bu})dv \\
&= ae^{-(au+bu)} + be^{-(au+bu)} = (a+b)e^{-(a+b)u}.
\end{aligned}
$$

That is, the minimum of two independent exponential random variables is exponentially distributed and its rate is the sum of the rates. Using distribution functions in 8.1, we had already seen this result. We now compute the density of V.

$$f_V(v) = \int_{u=0}^{v} (ae^{-au}be^{-bv}+ae^{-av}be^{-bu})du = b(1-e^{-av})e^{-bv}+a(1-e^{-bv})e^{-av}.$$

It is easy to see that the joint distribution of (U, V) is not the product of f_U and f_V. Therefore, U and V are not independent.

Example 11. Assume that X and Y are independent exponential random variables with rates a and b, respectively. Find the density of X/Y.
 We could set $U = X/Y$ and $V = X$, find the density of (U, V) and then find the density of U. However, in this case since exponential functions are easy to integrate we may use the distribution function technique of 8.1. Let $U = X/Y$. We have that

$$F_U(u) = P(U \le u) = P(X/Y \le u) = P(X \le uY).$$

Since X and Y are independent we know the joint density of (X, Y). Thus,

$$F_U(u) = \int_{x=0}^{\infty} \int_{y=x/u}^{\infty} ae^{-ax}be^{-by}dydx = \int_{x=0}^{\infty} ae^{-ax}e^{-bx/u}dx.$$

We get that

$$F_U(u) = \frac{a}{a + b/u} \text{ for } u > 0.$$

We now differentiate the distribution function to get the density of U:

$$f_U(u) = \frac{ab}{(au + b)^2} \text{ for } u > 0.$$

Exercises

1. Consider a uniform random vector on the triangle $\{(x, y) : 0 \le x \le y \le 1\}$.
   (a) Find the density of the vector (X, Y).
   (b) Find the marginal densities f_X and f_Y.
   (c) Are the two random variables X and Y independent?

2. Consider a uniform random vector on the square $\{(x, y) : 0 \le x \le 1;$ $0 \le y \le 1\}$.
   (a) Find the density of the vector (X, Y).
   (b) Find the marginal densities f_X and f_Y.
   (c) Are the two random variables X and Y independent?

3. Redo Example 5 assuming that the bus leaves at 7:03 precisely. What is the probability that I catch the bus?

4. Two friends have set an appointment between 8:00 and 8:30. Assume that the arrival time of each friend is uniformly distributed between 8:00 and 8:30. Assume also that the first that arrives waits for 15 minutes and then leaves. What is the probability that the friends miss each other?

5. Roll two dice. Let X be the sum and Y the minimum of the two dice.
   (a) Find the joint distribution of (X, Y).
   (b) Are X and Y independent?

6. Consider a uniform random vector on the triangle $\{(x, y) : 0 \le x \le y \le 1\}$. Find the correlation between X and Y.

7. Roll two dice. Let X be the sum and Y the minimum of the two dice. Find the correlation between X and Y.

8. Compute the correlation for the random variables of Example 6.

9. Let X and Y be two independent exponential random variables with rates λ and μ respectively. What is the probability that X is less than Y?

10. Show that if U has a Cauchy density, $f_U(u) = \frac{1}{\pi} \frac{1}{1+u^2}$, then $E(U)$ does not exist.

11. Let X and Y be two independent exponential random variables with rates λ.
   (a) Find the joint density of $(X + Y, X/Y)$.
   (b) Show that $X + Y$ and X/Y are independent.

12. Let X and Y be two exponential and independent random variables with rate a. Let $U = \min(X, Y)$, $V = \max(X, Y)$ and $D = V - U$.
   (a) Find the joint density of (U, D).
   (b) Are U and D independent?

13. Find the density of X/Y for X and Y where X and Y are two exponential independent random variables with rate a.

14. Let X and Y be two independent uniform random variables.
   (a) Find the density of XY.
   (b) Find the density of X/Y.

15. Let X and Y be two exponential and independent random variables with rate a. Let $U = X$ and $V = X + Y$. Find the joint density of (U, V).

8.3   Chi-square, Student Distributions and Normal Vectors

We start by constructing the Chi-square and Student distributions by using normal random variables. First, recall that a Gamma random variable with parameters $r > 0$ and $\lambda > 0$ has density

$$f(x) = \frac{\lambda^r}{\Gamma(r)} x^{r-1} e^{-\lambda x} \text{ for all } x > 0$$

where

$$\Gamma(r) = \int_0^\infty x^{r-1} e^{-x} dx.$$

The moment generating function of a Gamma random variable with parameters r and λ was computed in 7.1 and we found that

$$M_X(t) = (\frac{\lambda}{\lambda - t})^r \text{ for } t < \lambda.$$

Assume that $X_1, X_2, \ldots, X_n$ are independent Gamma random variables with parameters $(r_1, \lambda), (r_2, \lambda), \ldots, (r_n, \lambda)$, respectively. Then,

$$M_{X_1+X_2+\cdots+X_n}(t) = E\left(e^{t(X_1+X_2+\cdots+X_n)}\right) = E\left(e^{tX_1}\right) E\left(e^{tX_2}\right) \cdots E\left(e^{tX_n}\right)$$

where the last equality comes from the independence of the X_i. We now use the formula for the moment generating function of a Gamma to get

$$M_{X_1+X_2+\cdots+X_n}(t) = (\frac{\lambda}{\lambda - t})^{r_1} (\frac{\lambda}{\lambda - t})^{r_2} \cdots (\frac{\lambda}{\lambda - t})^{r_n} = (\frac{\lambda}{\lambda - t})^{r_1+r_2+\cdots+r_n}.$$

Recall that a moment generating function determines the distribution of a random variable. Therefore, the computation above shows that the sum of independent Gamma random variables with the same λ is a Gamma distribution.

Sum of Gamma Random Variables

Assume that $X_1, X_2, \ldots, X_n$ are independent Gamma random variables with parameters $(r_1, \lambda), (r_2, \lambda), \ldots, (r_n, \lambda)$, respectively (note that they all have the same λ). Then, $X_1 + X_2 + \cdots + X_n$ is a Gamma random variable with parameters $(r_1 + r_2 + \cdots + r_n, \lambda)$.

We use the preceding fact about Gamma distributions to show the following property of standard normal variables.

Chi-square Distribution

Assume that $Z_1, Z_2, \ldots, Z_n$ are independent standard normal distributions. Then,

$$X = Z_1^2 + Z_2^2 + \cdots + Z_n^2$$

is called a Chi-square random variable with n degrees of freedom. Its density is given by

$$f(x) = \frac{1}{2^{n/2}\Gamma(n/2)} x^{n/2-1} e^{-x/2} \text{ for } x > 0.$$

From Example 10 in 8.1 we know that the density of $Y = Z^2$ where Z is a standard normal variable is

$$f_Y(y) = \frac{1}{\sqrt{2\pi}} y^{-1/2} e^{-y/2} \text{ for } y \geq 0.$$

The density above is the density of a Chi-square random variable with one degree of freedom. This is consistent with the formula above since it is possible to show that $\Gamma(1/2) = \sqrt{\pi}$ (see Exercise 1 below). We also note that $\frac{1}{\sqrt{2\pi}} y^{-1/2} e^{-y/2}$ is the density of a Gamma with parameters 1/2 and 1/2. Hence,

$$X = Z_1^2 + Z_2^2 + \cdots + Z_n^2$$

is a sum of n independent Gamma random variables with parameters 1/2 and 1/2. Therefore, X is a Gamma random variable with parameters $r = n/2$ and $\lambda = 1/2$. The density of X is

$$f(x) = \frac{(1/2)^{n/2}}{\Gamma(n/2)} x^{n/2-1} e^{-x/2} \text{ for all } x > 0.$$

We saw in Chapter 5 that Chi-square distributions play a pivotal role in statistics. See Section 5.4 for sketches of the graph of Chi-square densities and for statistical applications. Another important distribution in statistics is the Student distribution.

Student Distribution

Let Z be a standard normal random variable and X be a Chi-square random variable with r degrees of freedom. Assume that X and Z are independent. Then,

$$T = \frac{Z}{\sqrt{X/r}}$$

is a Student random variable with r degrees of freedom. Its density is

$$f(t) = \frac{\Gamma(\frac{r+1}{2})}{\Gamma(r/2)\sqrt{\pi r}}(1 + t^2/r)^{-(r+1)/2}.$$

As r increases to infinity the density of a Student random variable with r degrees of freedom approaches the density of a standard normal distribution (see Section 5.3). We now find the density of a Student random variable with r degrees of freedom. Let

$$T = \frac{Z}{\sqrt{X/r}} \text{ and } U = X.$$

We invert the relations above to get

$$Z = T\sqrt{U/r} \text{ and } X = U.$$

The Jacobian of the transformation above is

$$\begin{vmatrix} \sqrt{u/r} & t/(2\sqrt{ur}) \\ 0 & 1 \end{vmatrix} = \sqrt{u/r}.$$

Since Z and X are independent, the density of (Z, X) is

$$\frac{1}{\sqrt{2\pi}\,2^{r/2}\Gamma(r/2)}e^{-z^2/2}x^{r/2-1}e^{-x/2} \text{ for any } z, x > 0.$$

The density of (T, U) is then

$$\frac{1}{\sqrt{2\pi}\,2^{r/2}\Gamma(r/2)}e^{-t^2u/(2r)}u^{r/2-1}e^{-u/2}\sqrt{u/r} \text{ for any } t, u > 0.$$

In order to get the density of T we integrate the joint density of (T, U) in u. We get

$$f_T(t) = \frac{1}{\sqrt{2\pi r}\,2^{r/2}\Gamma(r/2)}\int_0^\infty u^{\frac{r+1}{2}-1}e^{-u(\frac{t^2}{2r}+1/2)}du.$$

To compute the preceding integral, we use a Gamma density in the following way. We know that

$$\int_0^\infty \frac{\lambda^s}{\Gamma(s)}x^{s-1}e^{-\lambda x}dx = 1$$

for all $s > 0$ and $\lambda > 0$. This is so because the integrand above is the density of a Gamma random variable with parameters s and λ. Therefore,

$$\int_0^\infty x^{s-1} e^{-\lambda x} dx = \frac{\Gamma(s)}{\lambda^s}.$$

Now let $s = \frac{r+1}{2}$ and $\lambda = \frac{t^2}{2r} + 1/2$. Then we get

$$\int_0^\infty u^{\frac{r+1}{2}-1} e^{-u(\frac{t^2}{2r}+1/2)} du = \frac{\Gamma(\frac{r+1}{2})}{(\frac{t^2}{2r} + 1/2)^{\frac{r+1}{2}}}.$$

Thus,

$$f_T(t) = \frac{1}{\sqrt{2\pi r}\, 2^{r/2} \Gamma(r/2)} \frac{\Gamma(\frac{r+1}{2})}{(\frac{t^2}{2r} + 1/2)^{\frac{r+1}{2}}} = \frac{\Gamma(\frac{r+1}{2})}{\Gamma(r/2)\sqrt{\pi r}} (1 + t^2/r)^{-(r+1)/2}.$$

Normal random vectors

We now define normal vectors.

Normal Random Vectors

The vector $\mathbf{X} = \begin{pmatrix} X_1 \\ X_2 \\ \vdots \\ X_n \end{pmatrix}$ is said to be a normal random vector if there is an $n \times n$ matrix A, n independent standard normal variables $Z_1, Z_2, \ldots, Z_n$ and a constant vector $\mathbf{b}$ such that

$$\mathbf{X} = A\mathbf{Z} + \mathbf{b}$$

where $\mathbf{Z} = \begin{pmatrix} Z_1 \\ Z_2 \\ \vdots \\ Z_n \end{pmatrix}$.

Note that every X_i is a normal random variable. This is so because each X_i is a linear combination of the Z_i (plus the constant b_i) and since the Z_i are normal and independent X_i is a normal random variable as well (see Section 7.2).

In order to analyze some of the properties of normal random vectors we will use multivariate moment generating functions that we now define.

Multivariate Moment Generating Functions

Let $\mathbf{X}$ be a random vector. The moment generating function of $\mathbf{X}$ is defined as

$$M_{\mathbf{X}}(t_1, t_2, \ldots, t_n) = E(\exp(t_1 X_1 + t_2 X_2 + \cdots + t_n X_n)).$$

Multivariate moment generating functions have the two following important properties.

P1 Assume that

$$M_{\mathbf{X}}(t_1, t_2, \ldots, t_n) = M_{\mathbf{Y}}(t_1, t_2, \ldots, t_n) \text{ for } (t_1, t_2, \ldots, t_n) \text{ in } [-r, r]^n$$

for some $r > 0$; then the random vectors $\mathbf{X}$ and $\mathbf{Y}$ have the same distribution.

Property (P1) tells us the moment generating function completely characterizes the distribution of a random vector.

P2 The random variables $X_1, X_2, \ldots, X_n$ are independent if and only if

$$M_{\mathbf{X}}(t_1, t_2, \ldots, t_n) = M_{X_1}(t_1) M_{X_2}(t_2) \ldots M_{X_n}(t_n) \text{ for } (t_1, t_2, \ldots, t_n) \text{ in } [-r, r]^n$$

for some $r > 0$.

That is, in order for $X_1, X_2, \ldots, X_n$ to be independent, it is necessary and sufficient that the moment generating function of the vector $(X_1, X_2, \ldots, X_n)$ be the product of the moment generating functions of the random variables $X_1, X_2, \ldots, X_n$.

The proof of (P1) involves some advanced mathematics and we will skip it. However, assuming (P1) it is easy to prove (P2) and we will now do that. First, assume that the random variables $X_1, X_2, \ldots, X_n$ are independent. The density of the vector $\mathbf{X}$ is $f_{X_1}(x_1) f_{X_2}(x_2) \ldots f_{X_n}(x_n)$. Therefore,

$$M_{\mathbf{X}}(t_1, t_2, \ldots, t_n) =$$

$$\int_{x_1=-\infty}^{\infty} \int_{x_2=-\infty}^{\infty} \cdots \int_{x_n=-\infty}^{\infty} \exp(t_1 x_1 + \cdots + t_n x_n) f_{X_1}(x_1) \ldots f_{X_n}(x_n) dx_1 \ldots dx_n.$$

Since,

$$\exp(t_1 x_1 + t_2 x_2 + \cdots + t_n x_n) = \exp(t_1 x_1) \exp(t_2 x_2) \ldots \exp(t_n x_n),$$

we use Fubini's Theorem to get

$$M_{\mathbf{X}}(t_1, t_2, \ldots, t_n) = M_{X_1}(t_1) M_{X_2}(t_2) \ldots M_{X_n}(t_n).$$

We now prove the converse. Assume that the moment generating function of $\mathbf{X}$ is the product of the generating functions of the variables $X_1, X_2, \ldots, X_n$. Then $M_{\mathbf{X}}$ is equal to the moment generating function of a vector whose components $X_1, X_2, \ldots, X_n$ are independent. By using (P1) we see that the random variables $X_1, X_2, \ldots, X_n$ are independent. This completes the proof of (P2).

Before computing the moment generating function of a random vector we need to know how to compute the variance of a linear combination of random variables. Recall from Section 8.2 that

$$Var(X + Y) = Var(X) + Var(Y) + 2Cov(X, Y).$$

We may generalize the preceding to

$$Var\left(\sum_{i=1}^{n} t_i X_i\right) = \sum_{i=1}^{n} t_i^2 Var(X_i) + \sum_{i=1}^{n}\sum_{j\neq i} t_i t_j Cov(X_i, X_j).$$

By observing that $Var(X) = Cov(X, X)$ we get that

$$Var\left(\sum_{i=1}^{n} t_i X_i\right) = \sum_{i=1}^{n}\sum_{j=1}^{n} t_i t_j Cov(X_i, X_j).$$

We will now compute the joint moment generating function of a normal vector. First recall two important properties of normal random variables from Section 7.2.

Properties of Normal Random Variables

Let X be a normal random variable with mean μ and standard deviation σ; then the moment generating function of X is

$$M_X(t) = E(e^{tX}) = \exp(\mu t + \sigma^2/2).$$

Let $X_1, X_2, \ldots, X_n$ be a sequence of independent normal random variables and let $t_1, t_2, \ldots, t_n$ be a sequence of real numbers; then the linear combination

$$\sum_{i=1}^{n} t_i X_i$$

is also a normal random variable.

Assume that $\mathbf{X}$ is a normal vector. Then each X_i is a linear combination of independent standard normal variables. Note that a linear combination of linear combinations is also a linear combination. Therefore, $\sum_{i=1}^{n} t_i X_i$ is a linear combination of independent standard normal random variables as well. Hence, $L = \sum_{i=1}^{n} t_i X_i$ is a normal random variable. We compute the moment generating function of L.

$$M_L(s) = E(e^{sL}) = \exp(E(L)s + Var(L)s^2/2).$$

Since

$$E(L) = \sum_{i=1}^{n} t_i E(X_i) \text{ and } Var(L) = \sum_{i=1}^{n}\sum_{j=1}^{n} t_i t_j Cov(X_i, X_j),$$

we get that

$$M_L(s) = \exp\left(s\sum_{i=1}^{n} t_i E(X_i) + s^2\sum_{i=1}^{n}\sum_{j=1}^{n} t_i t_j Cov(X_i, X_j)/2\right).$$

There is a simple relation between the moment generating function of the vector $\mathbf{X}$ at $(t_1, t_2, \ldots, t_n)$ and the moment generating function of L.

$$M_{\mathbf{X}}(t_1, t_2, \ldots, t_n) = E(\exp(t_1 X_1 + t_2 X_2 + \cdots + t_n X_n)) = E(\exp(L)) = M_L(1).$$

We set $s = 1$ in the formula for $M_L(s)$ to get

$$M_{\mathbf{X}}(t_1, t_2, \ldots, t_n) = \exp\left(\sum_{i=1}^{n} t_i E(X_i) + \sum_{i=1}^{n}\sum_{j=1}^{n} t_i t_j Cov(X_i, X_j)/2\right).$$

The Moment Generating Function of a Normal Vector

Let $\mathbf{X} = \begin{pmatrix} X_1 \\ X_2 \\ \vdots \\ X_n \end{pmatrix}$ be a normal vector; then the moment generating function of $\mathbf{X}$ is

$$M_{\mathbf{X}}(t_1, t_2, \ldots, t_n) = \exp\left(\sum_{i=1}^{n} t_i E(X_i) + \sum_{i=1}^{n}\sum_{j=1}^{n} t_i t_j Cov(X_i, X_j)/2\right).$$

Example 1. Show that if the random variables $X_1, X_2, \ldots, X_n$ are independent and normally distributed, then the vector $\mathbf{X} = \begin{pmatrix} X_1 \\ X_2 \\ \vdots \\ X_n \end{pmatrix}$ is normally distributed.

We have at least two ways to show this. We may find a matrix A and a vector $\mathbf{b}$ such that $\mathbf{X} = A\mathbf{Z} + \mathbf{b}$, where $\mathbf{Z}$ is a vector whose components are independent standard normal random variables (see Exercise 7 below) or we may use moment generating functions. We use moment generating functions. Since the X_i are assumed to be independent we have by property (P2)

$$M_{\mathbf{X}}(t_1, t_2, \ldots, t_n) = M_{X_1}(t_1) M_{X_2}(t_2) \ldots M_{X_n}(t_n).$$

Since the X_i are normally distributed, we have for each $i = 1, 2, \ldots, n$,

$$M_{X_i}(s) = \exp(\mu_i s + s^2 \sigma_i^2/2).$$

Thus,

$$\begin{aligned}
&M_{\mathbf{X}}(t_1, t_2, \ldots, t_n) \\
&\quad = \exp(\mu_1 t_1 + t_1^2 \sigma_1^2/2) \exp(\mu_2 t_2 + t_2^2 \sigma_2^2/2) \ldots \exp(t_n \mu_n t_n + t_n^2 \sigma_n^2/2) \\
&\quad = \exp\left(\sum_{i=1}^{n} t_i \mu_i + \sum_{i=1}^{n} t_i^2 \sigma_i^2/2\right).
\end{aligned}$$

This is the moment generating function of a normal vector. By property (P1) this proves that **X** is a normal vector.

An easy consequence of the form of the moment generating function for a normal vector is the following property.

Independence and Covariance

Assume that $\mathbf{X} = \begin{pmatrix} X_1 \\ X_2 \end{pmatrix}$ is a normal random vector. Then X_1 and X_2 are independent if and only if

$$Cov(X_1, X_2) = 0.$$

We know already that if X_1 and X_2 are independent, then $Cov(X_1, X_2) = 0$. What is remarkable here is that the converse holds for normal random vectors. Assume that **X** is a normal random vector such that $Cov(X_1, X_2) = 0$. Therefore, the moment generating function of **X** is

$$M_{\mathbf{X}}(t_1, t_2) = \exp(t_1 E(X_1) + t_2 E(X_2) + t_1^2 Var(X_1)/2 + t_2^2 Var(X_2)/2).$$

This can be written as

$$\begin{aligned} M_{\mathbf{X}}(t_1, t_2) &= \exp(t_1 E(X_1) + t_1^2 Var(X_1)/2) \exp(t_2 E(X_2) + t_2^2 Var(X_2)/2) \\ &= M_{X_1}(t_1) M_{X_2}(t_2). \end{aligned}$$

By property (P2) this proves that X_1 and X_2 are independent.

Note that in order to use the property above one must first prove that **X** is a normal vector. Showing that the marginal densities are normal is not enough. Exercise 9 below gives an example of two normal random variables whose covariance is 0 and that are not independent. This is so because although the marginal densities are normal the joint density is not.

The joint distribution of the sample mean and variance in a normal sample

Let $X_1, X_2, \ldots, X_n$ be independent, normally distributed with mean μ and standard deviation σ. As we have seen in Section 5.1,

$$\bar{X} = \frac{X_1 + X_2 + \cdots + X_n}{n}$$

is an unbiased estimator of the mean μ and

$$S^2 = \frac{1}{n-1} \sum_{i=1}^{n} (X_i - \bar{X})^2$$

is an unbiased estimator of σ^2. That is,

$$E(\bar{X}) = \mu \text{ and } E(S^2) = \sigma^2.$$

$\bar{X}$ and S^2 are called the sample mean and the sample variance of the sample $X_1, \ldots,$ X_n. We will show that

$$\frac{\bar{X} - \mu}{S/\sqrt{n}}$$

follows a Student distribution with $n - 1$ degrees of freedom. This is an important result in statistics that we have used in Section 5.3. Let

$$\mathbf{X} = \begin{pmatrix} X_1 \\ X_2 \\ . \\ . \\ . \\ X_n \end{pmatrix} \quad \mathbf{D} = \begin{pmatrix} \bar{X} \\ X_1 - \bar{X} \\ X_2 - \bar{X} \\ . \\ X_n - \bar{X} \end{pmatrix}.$$

Note that

$$\mathbf{D} = \begin{pmatrix} 1/n & 1/n & . & . & . & 1/n \\ 1 - 1/n & -1/n & . & . & . & -1/n \\ -1/n & 1 - 1/n & . & . & . & -1/n \\ . & . & . & . & . \\ -1/n & -1/n & . & . & . & 1 - 1/n \end{pmatrix} \mathbf{X}.$$

According to Example 1, $\mathbf{X}$ is a normal random vector. Thus, $\mathbf{D}$ is the image by a linear transformation of a normal vector. It is easy to see that $\mathbf{D}$ is a normal random vector as well (see Exercise 4 below). We now compute for $i = 1, \ldots, n$,

$$Cov(\bar{X}, X_i - \bar{X}) = Cov(\bar{X}, X_i) - Cov(\bar{X}, \bar{X}).$$

By using the independence of the X_i it is easy to see that (see Section 4.1)

$$Cov(\bar{X}, \bar{X}) = Var(\bar{X}) = \sigma^2/n.$$

We now turn to

$$Cov(\bar{X}, X_i) = \frac{1}{n} \sum_{j=1}^{n} Cov(X_j, X_i).$$

Note that $Cov(X_j, X_i) = 0$ for $i \neq j$ since X_i and X_j are independent. Thus,

$$Cov(\bar{X}, X_i) = \frac{1}{n} Cov(X_i, X_i) = \sigma^2/n.$$

Therefore,

$$Cov(\bar{X}, X_i - \bar{X}) = 0$$

and since $\mathbf{D}$ is a normal random vector this is enough to show that, for every $i = 1, \ldots n$, $\bar{X}$ and $X_i - \bar{X}$ are independent. Since S^2 depends only on the differences $X_i - \bar{X}$ we have proved the following.

The Sample Mean and Sample Variance are Independent

Let $X_1, X_2, \ldots, X_n$ be independent, NORMALLY distributed with mean μ and standard deviation σ. Then, the sample mean

$$\bar{X} = \frac{X_1 + X_2 + \cdots + X_n}{n}$$

and the sample variance

$$S^2 = \frac{1}{n-1} \sum_{i=1}^{n} (X_i - \bar{X})^2$$

are independent.

This independence result relies heavily on the normality of the sample.

We know from Section 7.2 that a linear combination of independent normal random variables is also a normal variable. Therefore, $\bar{X}$ is normally distributed with mean μ and variance σ^2/n. We now turn to the distribution of S^2. We start with

$$\sum_{i=1}^{n} (X_i - \bar{X})^2 = \sum_{i=1}^{n} (X_i - \mu + \mu - \bar{X})^2$$

$$= \sum_{i=1}^{n} (X_i - \mu)^2 + 2\sum_{i=1}^{n} (X_i - \mu)(\mu - \bar{X}) + \sum_{i=1}^{n} (\mu - \bar{X})^2.$$

Note that

$$\sum_{i=1}^{n} (X_i - \mu)(\mu - \bar{X}) = (\mu - \bar{X})(n\bar{X} - n\mu) = -n(\mu - \bar{X})^2.$$

Thus,

$$(n-1)S^2 = \sum_{i=1}^{n} (X_i - \bar{X})^2 = \sum_{i=1}^{n} (X_i - \mu)^2 - n(\mu - \bar{X})^2.$$

We divide the preceding equality by σ^2 to get

$$(n-1)S^2/\sigma^2 = \sum_{i=1}^{n} \left(\frac{X_i - \mu}{\sigma}\right)^2 - n\left(\frac{\mu - \bar{X}}{\sigma}\right)^2.$$

The random variable

$$\sum_{i=1}^{n}\left(\frac{X_i - \mu}{\sigma}\right)^2$$

is the sum of the squares of n independent standard normal random variables. Thus, it is a Chi-square random variable with n degrees of freedom. On the other hand

$$n\left(\frac{\mu - \bar{X}}{\sigma}\right)^2 = \left(\frac{\bar{X} - \mu}{\sigma/\sqrt{n}}\right)^2$$

is the square of a standard normal random variable (since the expected value of $\bar{X}$ is μ and its standard deviation is $\sigma/\sqrt{n}$). So it is a Chi-square random variable with one degree of freedom. We are now going to find the distribution of S^2 by using moment generating functions. We rewrite the identity

$$(n-1)S^2/\sigma^2 = \sum_{i=1}^{n}\left(\frac{X_i - \mu}{\sigma}\right)^2 - n\left(\frac{\mu - \bar{X}}{\sigma}\right)^2$$

as

$$(n-1)S^2/\sigma^2 + n\left(\frac{\mu - \bar{X}}{\sigma}\right)^2 = \sum_{i=1}^{n}\left(\frac{X_i - \mu}{\sigma}\right)^2.$$

We compute the moment generating functions:

$$E\left[\exp\left(t(n-1)S^2/\sigma^2 + tn\left(\frac{\mu - \bar{X}}{\sigma}\right)^2\right)\right] = E\left[\exp\left(t\sum_{i=1}^{n}\left(\frac{X_i - \mu}{\sigma}\right)^2\right)\right].$$

Recall that a Chi-square random variable with k degrees of freedom has a moment generating function $(1 - 2t)^{-k/2}$. We use also that S^2 and $\bar{X}$ are independent to get

$$E[\exp(t(n-1)S^2/\sigma^2)](1 - 2t)^{-1/2} = (1 - 2t)^{-n/2}.$$

Thus,

$$E[\exp(t(n-1)S^2/\sigma^2)] = (1 - 2t)^{-(n-1)/2}.$$

That is, $(n-1)S^2/\sigma^2$ follows a Chi-square distribution with $n - 1$ degrees of freedom.

Finally, since

$$\frac{\bar{X} - \mu}{\sigma/\sqrt{n}}$$

is a standard normal random variable that is independent of the Chi-square random variable with $n - 1$ degrees of freedom $(n - 1)S^2/\sigma^2$, we have that

$$\frac{\frac{\bar{X}-\mu}{\sigma/\sqrt{n}}}{\sqrt{\frac{(n-1)S^2}{(n-1)\sigma^2}}} = \frac{\bar{X} - \mu}{S/\sqrt{n}}$$

follows a Student distribution with $n - 1$ degrees of freedom. We now summarize our results.

Joint Distribution of the Sample Mean and the Sample Variance

Let $X_1, X_2, \ldots, X_n$ be independent, NORMALLY distributed with mean μ and standard deviation σ. Then $\bar{X}$ and S^2 are independent. Moreover,

$$(n - 1)S^2/\sigma^2$$

follows a Chi-square distribution with $n - 1$ degrees of freedom. Finally,

$$\frac{\bar{X} - \mu}{S/\sqrt{n}}$$

is a Student random variable with $n - 1$ degrees of freedom.

Exercises

1. (a) Show that if $C_1 g(x)$ and $C_2 g(x)$ are both density functions, then $C_1 = C_2$.

(b) Use (a) to compare the density of Z^2 (given by Example 10 in Section 8.1) to the density of a Gamma random variable with parameters 1/2 and 1/2 to conclude that $\Gamma(1/2) = \sqrt{\pi}$.

2. (a) Assume that X and Y are independent Chi-square random variables with degrees of freedom n and m respectively. Show that $X + Y$ is also a Chi-square random variable with $n + m$ degrees of freedom.

(b) Assume that X and Y are independent and that X and $X + Y$ are Chi-square random variables with degrees of freedom n and m respectively. Show that Y is also a Chi-square random variable with $m - n$ degrees of freedom.

3. Use a Gamma density to compute the integral

$$\int_0^\infty x^5 e^{-2x} dx.$$

4. Let $\mathbf{X}$ be a normal vector, let B be a matrix. Show that $\mathbf{Y} = B\mathbf{X}$ is also a normal vector.

5. Assume that $Var(X) = 1$, $Var(Y) = 2$ and $Cov(X, Y) = -1$. Compute $Cov(X - 2Y, X + Y)$.

6. Assume that (X, Y) is a normal vector with $Var(X) = 1$, $Var(Y) = 2$ and $Cov(X, Y) = -1$. Find the moment generating function of the vector (X, Y).

7. Assume that X_1 and X_2 are independent normally distributed random variables with means μ_1, μ_2 and standard deviations σ_1, σ_2 respectively. Show that $\mathbf{X} = \begin{pmatrix} X_1 \\ X_2 \end{pmatrix}$ is a normal vector by finding a matrix A and a vector $\mathbf{b}$ such that

$$\mathbf{X} = \mathbf{AZ} + \mathbf{b}$$

where $\mathbf{Z} = \begin{pmatrix} Z_1 \\ Z_2 \end{pmatrix}$ and Z_1 and Z_2 are independent standard normal random variables.

8. Assume that X_1 and X_2 are independent normally distributed random variables with mean μ and standard deviation σ.
   (a) Compute the $Cov(X_1 - X_2, X_1 + X_2)$.
   (b) Are $X_1 - X_2$ and $X_1 + X_2$ independent?

9. In this exercise we will construct two normal random variables X and Y such that $Cov(X, Y) = 0$ and such that X and Y are not independent. Let

$$n(x, y) = \frac{1}{2\pi} \exp(-(x^2 + y^2)/2).$$

That is, n is the joint density of two independent standard normal random variables. Let D_1, D_2, D_3 and D_4 be the interiors of the circles with radius 1 and centered at $(2, 2)$, $(-2, 2)$, $(-2, -2)$ and $(2, -2)$, respectively. Define

$$\begin{aligned}
f(x, y) &= n(x, y) \text{ for } (x, y) \text{ not in } D_1 \cup D_2 \cup D_3 \cup D_4 \\
f(x, y) &= n(x, y) + m \text{ for } (x, y) \text{ in } D_1 \\
f(x, y) &= n(x, y) - m \text{ for } (x, y) \text{ in } D_2 \\
f(x, y) &= n(x, y) + m \text{ for } (x, y) \text{ in } D_3 \\
f(x, y) &= n(x, y) - m \text{ for } (x, y) \text{ in } D_4
\end{aligned}$$

where m is a constant small enough so that $f(x, y)$ is always strictly positive.
   (a) Show that f is a density.
   (b) Show that X and Y are standard normal random variables.
   (c) Compute the covariance of X and Y.
   (d) Show that X and Y are not independent.

10. Assume that (X, Y) is a normal vector with $Var(X) = 1$, $Var(Y) = 2$ and $Cov(X, Y) = -1$. Find a so that X and $X + aY$ are independent.

Appendices

Further Reading

Probability

The following two references are at a slightly higher level than this book. They cover additional topics and examples in probability. They are:

The Essentials of Probability by R. Durrett, The Duxbury Press, 1994.

Probability by J. Pitman, Springer-Verlag, New York, 1993.

An Introduction to Probability Theory and its Applications by W. Feller, volume I, third edition, Wiley, 1971, is at a substantially higher level than this book. It has influenced several generations of probabilists and covers hundreds of interesting topics. It is a GREAT book.

Statistics

A very good elementary introduction to statistics is *Introduction to the Practice of Statistics* by D. Moore and G. McCabe, third edition, Freeman, 1999. A more mathematical approach to statistics is contained in *Probability and Statistics* by K. Hastings, Addison-Wesley, 1997.

Normal Table

Table 1. The table below gives $P(0 < Z < z)$ for a standard normal random variable Z. For instance $P(0 < Z < .43) = .1664$.

z	.00	.01	.02	.03	.04	.05	.06	.07	.08	.09
.0	.0000	.0040	.0080	.0120	.0160	.0199	.0239	.0279	.0319	.0359
.1	.0398	.0438	.0478	.0517	.0557	.0596	.0636	.0675	.0714	.0753
.2	.0793	.0832	.0871	.0910	.0948	.0987	.1026	.1064	.1103	.1141
.3	.1179	.1217	.1255	.1293	.1331	.1368	.1406	.1443	.1480	.1517
.4	.1554	.1591	.1628	.1664	.1700	.1736	.1772	.1808	.1844	.1879
.5	.1915	.1950	.1985	.2019	.2054	.2088	.2123	.2157	.2190	.2224
.6	.2257	.2291	.2324	.2357	.2389	.2422	.2454	.2486	.2517	.2549
.7	.2580	.2611	.2642	.2673	.2704	.2734	.2764	.2794	.2823	.2852
.8	.2881	.2910	.2939	.2967	.2995	.3023	.3051	.3078	.3106	.3133
.9	.3159	.3186	.3212	.3238	.3264	.3289	.3315	.3340	.3365	.3389
1.0	.3413	.3438	.3461	.3485	.3508	.3531	.3554	.3577	.3599	.3621
1.1	.3643	.3665	.3686	.3708	.3729	.3749	.3770	.3790	.3810	.3830
1.2	.3849	.3869	.3888	.3907	.3925	.3944	.3962	.3980	.3997	.4015
1.3	.4032	.4049	.4066	.4082	.4099	.4115	.4131	.4147	.4162	.4177
1.4	.4192	.4207	.4222	.4236	.4251	.4265	.4279	.4292	.4306	.4319
1.5	.4332	.4345	.4357	.4370	.4382	.4394	.4406	.4418	.4429	.4441
1.6	.4552	.4463	.4474	.4484	.4495	.4505	.4515	.4525	.4535	.4545
1.7	.4554	.4564	.4573	.4582	.4591	.4599	.4608	.4616	.4625	.4633
1.8	.4641	.4649	.4656	.4664	.4671	.4678	.4686	.4693	.4699	.4706
1.9	.4713	.4719	.4726	.4732	.4738	.4744	.4750	.4756	.4761	.4767
2.0	.4772	.4778	.4783	.4788	.4793	.4798	.4803	.4808	.4812	.4817
2.1	.4821	.4826	.4830	.4834	.4838	.4842	.4846	.4850	.4854	.4857
2.2	.4861	.4864	.4868	.4871	.4875	.4878	.4881	.4884	.4887	.4890
2.3	.4893	.4896	.4898	.4901	.4904	.4906	.4909	.4911	.4913	.4916
2.4	.4918	.4920	.4922	.4925	.4927	.4929	.4931	.4932	.4934	.4936
2.5	.4938	.4940	.4941	.4943	.4945	.4946	.4948	.4949	.4951	.4952
2.6	.4953	.4955	.4956	.4957	.4959	.4960	.4961	.4962	.4963	.4964
2.7	.4965	.4966	.4967	.4968	.4969	.4970	.4971	.4972	.4973	.4974
2.8	.4974	.4975	.4976	.4977	.4977	.4978	.4979	.4979	.4980	.4981
2.9	.4981	.4982	.4982	.4983	.4984	.4984	.4985	.4985	.4986	.4986
3.0	.4987	.4987	.4987	.4988	.4988	.4989	.4989	.4989	.4990	.4990

Student Table

Table 2. The table below gives t_a such that $P(|t(n)| < t_a) = a$ where $t(n)$ is a Student distribution with n degrees of freedom. For instance, we read that $P(|t(5)| < 1.48) = 0.8$.

n	$a = 0.6$	$a = 0.7$	$a = 0.8$	$a = 0.9$	$a = 0.95$
1	1.38	1.96	3.08	6.31	12.71
2	1.06	1.39	1.89	2.92	4.30
3	0.98	1.25	1.64	2.35	3.18
4	0.94	1.19	1.53	2.13	2.78
5	0.92	1.16	1.48	2.02	2.57
6	0.91	1.13	1.44	1.94	2.45
7	0.90	1.12	1.41	1.89	2.36
8	0.89	1.11	1.40	1.86	2.31
9	0.88	1.10	1.38	1.83	2.26
10	0.88	1.09	1.37	1.81	2.23
11	0.88	1.09	1.36	1.80	2.20
12	0.87	1.08	1.36	1.78	2.18
13	0.87	1.08	1.35	1.77	2.16
14	0.87	1.08	1.35	1.76	2.14
15	0.87	1.07	1.34	1.75	2.13
16	0.86	1.07	1.34	1.75	2.12
17	0.86	1.07	1.33	1.74	2.11
18	0.86	1.07	1.33	1.73	2.10
19	0.86	1.07	1.33	1.73	2.09
20	0.86	1.06	1.33	1.72	2.09
21	0.86	1.06	1.32	1.72	2.08
22	0.86	1.06	1.32	1.72	2.07
23	0.86	1.06	1.32	1.71	2.07
24	0.86	1.06	1.32	1.71	2.06
25	0.86	1.06	1.32	1.71	2.06
∞	0.84	1.04	1.28	1.64	2.01

Chi-Square Table

Table 3. The table below gives χ_a such that $P(\chi(n) < \chi_a) = a$ where $\chi(n)$ is a Chi-Square distribution with n degrees of freedom. For instance, we read that $P(\chi(6) < 1.64) = 0.05$.

n	$a = 0.01$	$a = 0.05$	$a = 0.90$	$a = 0.95$	$a = 0.99$
1	0.00	0.00	2.71	3.84	6.63
2	0.02	0.10	4.61	5.99	9.21
3	0.11	0.35	6.25	7.81	11.34
4	0.30	0.71	7.78	9.49	13.28
5	0.55	1.15	9.24	11.07	15.09
6	0.87	1.64	10.64	12.59	16.81
7	1.24	2.17	12.02	14.07	18.48
8	1.65	2.73	13.36	15.51	20.09
9	2.09	3.33	14.68	16.92	21.67
10	2.56	3.94	15.99	18.31	23.21
11	3.05	4.57	17.28	19.68	24.72
12	3.57	5.23	18.55	21.03	26.22
13	4.11	5.89	19.81	22.36	27.69
14	4.66	6.57	21.06	23.68	29.14
15	5.23	7.26	22.31	25.00	30.58
16	5.81	7.96	23.54	26.30	32.00
17	6.41	8.67	24.77	27.59	33.41
18	7.01	9.39	25.99	28.87	34.81
19	7.63	10.12	27.20	30.14	36.19
20	8.26	10.85	28.41	31.41	37.57
21	8.90	11.59	29.62	32.67	38.93
22	9.54	12.34	30.81	33.92	40.29
23	10.20	13.09	32.01	35.17	41.64
24	10.86	13.85	33.20	36.42	42.98
25	11.52	14.61	34.38	37.65	44.31

Index

Also by the author

Classical and Spatial Stochastic Processes, 1999
ISBN 0-8176-4081-9, Hardcover, 192 pages, $54.95

An appetizing textbook for a first course in stochastic processes. It guides the reader in a very clever manner from classical ideas to some of the most interesting modern results... All essential facts are presented with clear proofs, illustrated by beautiful examples... The book is well organized, has informative chapter summaries, and presents interesting exercises. The clear proofs are concentrated at the ends of the chapters making it easy to find the results. The style is a good balance of mathematical rigorosity and user-friendly explanation.

—Biometric Journal

This small book is well-written and well-organized... Only simple results are treated...but at the same time many ideas needed for more complicated cases are hidden and in fact very close. The second part is a really elementary introduction to the area of spatial processes... All sections are easily readable and it is rather tentative for the reviewer to learn them more deeply by organizing a course based on this book. The reader can be really surprised seeing how simple the lectures on these complicated topics can be. At the same time such important questions as phase transitions and their properties for some models and the estimates for certain critical values are discussed rigorously... This is indeed a first course on stochastic processes and also a masterful introduction to some modern chapters of the theory.

—Zentralblatt für Mathematik